DoingBiology

Joel B. Hagen
Radford University

Douglas Allchin
University of Texas at El Paso

Fred Singer
Radford University

HarperCollinsCollegePublishers

*To Sue for her good-natured support and
encouragement throughout this project.—J.B.H.*

*To Pamela Goss Cook, who began it all with a homework problem in 10th-grade
chemistry: Never underestimate the value of a high school teacher.—D.A.*

*To Cindy, Jed, and Alison for their curiosity about life and companionship
throughout life.—F.S.*

Sponsoring Editor: Liz Covello
Project Coordination: Elm Street Publishing Services, Inc.
Art Development Editor: Vita Jay
Design Administrator: Jess Schaal
Text and Cover Design: Julie Anderson
Cover Illustration/Photo: Dr. Dennis Kunkel/Phototake
Photo Research: Leslie Coopersmith
Production Administrator: Randee Wire
Compositor: Elm Street Publishing Services, Inc.
Printer and Binder: R. R. Donnelley & Sons Company
Cover Printer: Lehigh Press Lithographers, Inc.

Doing Biology
Copyright © 1996 by HarperCollins College Publishers

Library of Congress Cataloging-in-Publication Data

Hagen, Joel Bartholemew.
  Doing biology / Joel Hagen, Douglas Allchin, Fred Singer.
    p. cm.
  Includes bibliographical references.
  ISBN 0-673-99638-7
  1. Biology 2. Biology—Case studies. I. Allchin, Douglas, 1956– . II Singer, Fred, 1952– . III.
  Title.
QH308.2.H335 1996
574—dc20   96-4220

96 97 98 99 9 8 7 6 5 4 3 2

Contents

Preface

Those ignorant of the historical development of science are not likely ever to understand fully the nature of science and scientific research.

—*Hans Krebs, Nobel laureate*

Every biological fact has a story behind it. The story helps us understand how the fact is justified. It also tells us how science is done. If we do not know this, we remain ignorant of how scientists work and how important scientific discoveries have been made. Krebs's own discovery of the metabolic cycle that bears his name is an excellent example of scientific problem solving. It illustrates the creative processes by which our knowledge of the living world is generated, and the critical processes by which scientific claims are evaluated. Like many other case studies, the discovery of the Krebs cycle is also a compelling story of personal triumph over adversity.

Without understanding the background of discoveries we have little to justify our scientific beliefs. We must rely on the authority of scientists themselves. Authorities are often wrong, however. A stronger foundation for knowledge comes from examining its justification: the alternative explanations, the various forms of evidence, and the reasoning strategies used to draw conclusions. This allows us to think critically about both the process and content of science. We can analyze important questions dealing with scientific and social issues such as medicine, the environment, the teaching of evolution in public schools, the ethics of genetic research—and other debates that will arise in the future. This understanding cannot be gained simply by learning currently accepted facts and theories. We must also understand how biologists *do biology*.

We have adopted a historical perspective, but this book is not really about history. Our objective is to look at the past to better understand biology today. All our case studies deal with or touch upon biological topics included in a typical introductory biology course. Although some of the scientists may be unfamiliar, their discoveries are not.

Historical perspectives are not new. Indeed, biology textbooks often celebrate discoveries from the distant past: van Leeuwenhoeck's pioneering microscopic investigations, for example, or van Helmont's intriguing experiments on plant growth. We have omitted these early examples of biology because we find them fraught with problems of interpretation. The scientific issues facing van Helmont and the culture of which he was a part were very different from today's. To really understand van Helmont's science, we would have to immerse ourselves in the ideas of the seventeenth century and regard seriously some concepts that today seem wildly implausible and *un*scientific. Were we to discuss his experiments in our present terms, however, we would capture only a shadow of van Helmont's actual achievement. By making early biologists appear more modern than they really were, we distort both history and the process of scientific discovery. For this reason, we have focused on episodes from the past century.

Still, even recent scientific discoveries look different now than they did in their original historical context. The solution to a problem appeared less obvious to the biologists who originally struggled with it than it does to us today. To show how someone made the transition to our current state of knowledge, we have reconstructed how biologists viewed the world *before* a discovery was made.

Taking this historical perspective is a useful approach to teaching because early scientific theories are often quite similar to the ideas of students who are struggling with a modern concept for the first time. In recreating historical moments of uncertainty and discovery, we hope to convey a sense of "science-in-the-making." We trust that teachers, too, will find it useful to place themselves in the shoes of famous biologists and to face historically significant problems and original data, forsaking the privilege of already knowing the answer. How did renowned biologists reach certain conclusions with the intellectual and technical resources available to them at the time of the discovery? In answering these questions, we hope to faithfully portray how scientific knowledge develops.

Toward these ends, we have adopted a case study approach. We believe that science is complex and richly textured. Practicing biologists know that scientists do not all share a single "cookbook" approach to research. What is often referred to as "*the* scientific method" is really a diverse "toolbox" of methods, including simple observation, indirect observation with instruments, inferences from historical evidence, controlled experiments, demonstrations, modeling and simulation, correlation studies, and reasoning by analogy. Each of these methods has unique strengths and limitations. Through case studies, we hope to highlight the diversity of approaches used by biologists in different fields.

The case study method is also particularly appropriate for presenting the human character of science. Most chapters focus on individual biologists. We hope to show how biologists' personalities and styles of investigation shaped the outcome of their biological research. We also wish to underscore biologists' deep commitment to research and to convey some of the excitement of *doing biology*. At the same time, we do not want to idealize or romanticize accounts of scientific discovery. Without dwelling unduly on character flaws, we present biologists as people with emotions, ambitions, and theoretical biases. After all, biologists are human, and these human characteristics also influence *doing biology*.

While focusing on individuals, we should always keep in mind that science is a social activity. Scientists both collaborate and compete within scientific communities. They are also influenced by the broader society of which they are members. We have placed each of our cases within a broader historical context, although this is not the central focus of the book. We encourage those readers interested in the cultural dimension of biology to explore the rich literature on this topic produced by social historians and sociologists.

In our selected case studies, we offer a wide cross-section for understanding science as an intellectual, practical, and social process. We do not intend to present encyclopedic coverage of the history of biology. Too many cases would simply make this book unwieldy. We chose historical episodes that are closely tied to topics in introductory biology courses, that exemplify important characteristics of scientific practice, and that are less familiar than those found in most textbooks.

We have also tried to strike a balance in the length of individual chapters. Textbooks often include short historical sketches that can be read in a minute or two. Such vignettes are valuable for introducing biological topics but are too superficial to show the creative process of discovery. We hope our accounts more fully reveal how biology is done but still remain short enough to read in one sitting. For those interested in greater depth of coverage, we provide a "Suggested Reading" list at the end of each chapter.

This volume will be a useful companion to any introductory biology text. Our problem-solving approach is designed to engage students in *doing biology,* not just reading about its conclusions. Each chapter provides an occasion for discussion—during recitation sessions, laboratory, or lecture. Alternatively, chapters may be assigned as supplementary reading. We have embedded problems within the text and "Questions and Activities" at the end of each chapter to serve as points of departure.

Ultimately, this book addresses a need expressed in several recent proposals for reforming science teaching. We have not written an alternative textbook, nor do we offer a new curriculum. We aim to complement existing teaching resources and to fill an obvious void. Textbooks teach biological content. We hope this volume goes a significant step further by helping students to learn about *doing biology*.

ACKNOWLEDGMENTS

This project grew out of a meeting of fifty historians and biology teachers held at Radford University in May 1993. We wish to thank all of the participants of the "Biology in Action" conference for their helpful ideas, criticisms, and discussions during workshop sessions on how to use historical case studies to teach biology. Financial support for this project was provided by grants from the National Science Foundation (USE-9150782), the State Council on Higher Education for Virginia, V-QUEST, Radford University, and the Radford University Foundation.

We are grateful to the following individuals for reading and commenting on various chapters in this book: John Beatty, Neil Campbell, Paul Farber, Steve Fifield, Roger Guillemin, Sue Hagen, Denise Kania, Rich Kliman, Chuck Kugler, Cindy Miller, Rich Murphy, Mary Roberts, Orion Rogers, David Rudge, Michael Shewmaker, Jed Singer, Sara Tjossem, and Chris Young. We wish to acknowledge additional advice and assistance from Gar Allen, Bud Bennett, Rhonda Goad, Vickie Goad, Verna Holoman, Franklin Jones, Pat Mikesell, Helen Mitchell, Judy Niehaus, Steve Pontius, John Prebble, Louis Pyenson, Peter Rich, Warren Self, Greg Zagursky, and Glynn Laboratories.

We also wish to thank the following reviewers of this text:

Robert Beckmann, *North Carolina State University*
Robert Bergad, *University of Minnesota*
Nathan Dubowsky, *Westchester Community College*
Paul Farber, *Oregon State University*
Maura Flannery, *St. John's University*
Florence Juillerat, *Indiana University*
Patti Soderberg, *Beloit College*
Robin Tyser, *University of Wisconsin*

We thank our editors—Liz Covello, Glynn Davies, Susan McLaughlin, Thom Moore, Ingrid Mount, Ed Moura, Kathy Richmond, and Bonnie Roesch for their support and encouragement. We are particularly grateful to Liz Covello for shepherding this project through the final stages of production.

Joel B. Hagen
Douglas Allchin
Fred Singer

CHAPTER 1

H. B. D. Kettlewell *&* the Peppered Moths

JOEL B. HAGEN

☐ *INTRODUCTION*

When he wrote his greatest book, *On the Origin of Species* (1859), Charles Darwin set out to accomplish two goals. First, he wanted to demonstrate that species change over time through the process of organic evolution. Second, he wanted to convince his readers that this evolutionary process occurs primarily as the result of natural selection. Eventually, both of these goals were achieved, but during his lifetime Darwin enjoyed only a partial victory. By the time of his death in 1881, virtually the entire scientific community had accepted the fact of evolution. But only a few biologists believed that natural selection was the primary cause of evolutionary change. For half a century after Darwin's death, biologists debated several alternative theories of evolution.

The eventual acceptance of natural selection depended heavily upon genetics, which provided a convincing explanation for the origin and spread of hereditary variations. Darwin realized the importance of heredity for natural selection, but he had no satisfactory explanation for it. By 1920 the basic principles of Mendelian genetics were well established. Combining the theory of natural selection with these new concepts of heredity, mathematical theorists demonstrated that evolution could occur as Darwin had claimed. This combination of Mendelism and Darwinism also caused the decline of alternative theories of evolution. By World War II, most biologists had rejected once-popular evolutionary ideas such as large-scale mutations or the inheritance of acquired traits. The Darwinian revolution was complete.

The evidence for natural selection was overwhelming, but most of it was indirect. The widespread acceptance of the theory, therefore, raised intriguing questions. Could biologists discover cases of populations evolving through natural selection? Darwin and many of his followers assumed that evolution was such a gradual process that it might take decades or centuries to detect changes in a population. On the other hand, some theoretical biologists predicted that populations could evolve very rapidly if natural selection was intense (see Chapter 15). Could suitable populations be found to test this prediction?

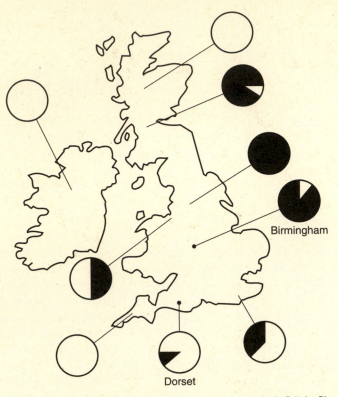

FIGURE 1.1   Some of the local populations of peppered moths in Britain. Pie diagrams show the relative frequency of dark- and light-winged individuals in each population. Birmingham and Dorset were the sites of H. B. D. Kettlewell's famous mark-release-recapture experiments.

THE PHENOMENON OF INDUSTRIAL MELANISM

Collecting insects is a popular hobby in England, and amateur naturalists were quick to record dramatic changes in wing coloration that occurred in several species of moths, notably the peppered moth (*Biston betularia*). This color change, which was due to a black pigment called melanin, seemed to occur most frequently in moths living near industrial cities (Figure 1.1). Populations made up almost entirely of light-winged individuals in 1800 had become mostly dark-winged a century later. Recognizing the evolutionary importance of this change, professional biologists also turned their attention to this phenomenon of **industrial melanism**.

In a widely publicized series of experiments conducted during the 1920s, the British entomologist J. W. Heslop Harrison fed caterpillars leaves coated with toxic compounds commonly found in soot. For example, in one experiment, caterpillars captured in a nonpolluted forest were fed leaves coated with lead nitrate.

After pupation, 53 light-winged moths and 3 dark-winged moths emerged from cocoons. All of the caterpillars in the control group, which were fed unpolluted leaves, developed into light-winged moths.

Harrison concluded that these results were due to mutations induced by chemical pollutants. Because the dark wings were inherited by many descendants of his experimental moths, Harrison also claimed that he had documented a case of inheritance of acquired traits. Publishing his experimental results in the prestigious British journal, *Nature*, Harrison presented his theory as a clear-cut alternative to natural selection.

PROBLEM

In his *Nature* article, Harrison did not provide many details about his experiments. List alternative explanations (other than mutation) that could also account for the unexpectedly large number of dark-winged moths.

Later attempts to replicate Harrison's experiments failed, and his explanation for industrial melanism was criticized by Darwinians. For example, the prominent theoretical biologist R. A. Fisher pointed out that Harrison's explanation required a mutation rate much higher than any previously reported. Nonetheless, Harrison continued to argue for his Lamarckian theory. Because he was a distinguished member of the British scientific community, his ideas could not simply be ignored. Therefore, this controversy set the stage for later research on industrial melanism.

A decade after Harrison published the results of his experiments, the geneticist E. B. Ford presented an alternative explanation. According to Ford, industrial melanism could be explained by natural selection acting on rare mutations. According to Ford, random mutations had always produced a few melanic moths in light-winged populations, but the mutants were quickly eliminated by natural selection. However, in polluted areas melanism proved adaptive, gained a selective advantage, and rapidly spread through the population.

You should note the important differences between the hypotheses proposed by Harrison and Ford. Harrison claimed that mutations occurred as a direct result of pollution and that they occurred simultaneously in many members of the population. Ford claimed that the genetic changes were not directly caused by pollution and that these mutations had always occurred *in very small numbers* in the population. According to Harrison, evolution occurred because many members of the population simultaneously mutated to the dark form. According to Ford, evolution resulted from the higher survival rate and reproductive success of rare mutants compared to their more common light-winged relatives.

Ford did not conduct experiments to test his hypothesis, and his explanation left a number of important questions unanswered. What exactly was the advantage of melanism? Was it really camouflage, or was it some unrelated physiological advantage linked to wing coloration? Could predators actually distinguish between melanic and nonmelanic moths? Could natural selection account for such a rapid increase in melanic individuals in polluted areas? The stage was set for a dramatic experimental test when H. B. D. Kettlewell began studying peppered moths in 1951.

H. B. D. KETTLEWELL AND ECOLOGICAL GENETICS

Henry Bernard Davis Kettlewell was the son of a successful businessman. As a boy, he attended prestigious private schools (so-called "public schools" in Britain), and he later studied both zoology and medicine at Cambridge University. For several years, he happily practiced medicine in a small town in southern England. Apparently because he disapproved of the nationalization of health care in Britain after World War II, Kettlewell left medical practice in 1949. For a time he worked on a locust control project in South Africa, later returning to England to join E. B. Ford's laboratory at Oxford University.

Kettlewell was an avid field biologist who loved adventure. When he left the locust control project, he drove from South Africa to Egypt, certainly not a journey for the fainthearted. As his biographer recalled, "He was a big man, with a personality larger than life. Kind, charming, and irascible, he had a huge and infectious ebullience and energy, could be the life and soul of any party, and was much loved by his friends."

When Kettlewell arrived, Oxford University was a bustling center of activity in field biology. Aside from Ford, the university was home to a number of other internationally famous naturalists, including the ornithologist David Lack, the ecologist Charles Elton, and the ethologist Niko Tinbergen. This scientific community proved valuable; Tinbergen later helped Kettlewell demonstrate that birds choose their prey based upon differences in wing color.

Ford and his associates were unusual geneticists because they preferred doing experiments in the field rather than in the laboratory. Working closely with the theoretical biologist R. A. Fisher, Ford had designed very sophisticated experimental methods for studying the genetics of natural populations, and unlike many naturalists at that time, he used statistics to analyze his experimental data. Partly influenced by Fisher's mathematical models, Ford believed that natural selection was the most important cause of evolution. According to Fisher and Ford, natural selection was often so intense that adaptations could quickly spread through a population if the environment changed. Ford's combination of experimental field studies, mathematical theorizing, and emphasis upon natural selection became known as "ecological genetics." It was an approach that Kettlewell skillfully applied to solve the problem of industrial melanism.

Kettlewell was convinced that wing coloration is an important adaptation in peppered moths. Females rarely fly; some spend their entire lives on a single branch. Males fly during the night and rest on tree trunks during the day. Both sexes rest with their wings open; therefore, camouflaged wings ought to provide some protection against predatory birds. Could this explain industrial melanism?

According to Kettlewell's hypothesis, light-colored wings provided camouflage when the moths rested on the lichens that often cover tree trunks in British forests. This cryptic coloration is so effective that moths are often invisible to humans standing only a few feet away (Figure 1.2). Melanic individuals are occasionally found in rural populations, but because their dark wings contrast with lichen-covered tree trunks, they are more likely to be eaten by predatory birds. As a result of the indus-

FIGURE 1.2 Dark- and light-winged varieties of the peppered moth resting on two contrasting backgrounds. *Source:* © M. W. F. Tweedie/Photo Researchers, Inc.

trial revolution, many areas of Britain became heavily polluted. Smoke from burning coal killed the lichens and caused trees to darken with soot. Here the adaptive value of wing coloration was reversed. Against a dark background, melanic moths are camouflaged and light-colored moths are conspicuous. At least, that is what Kettlewell thought. Many questions remained, however. Were birds really fooled by this camouflage? Did wing color really provide a significant advantage to camouflaged individuals? Most importantly, could Kettlewell design convincing experiments to confirm his hypotheses?

KETTLEWELL'S EARLY LABORATORY EXPERIMENTS

One of the first questions that Kettlewell set out to answer was whether moths choose the background on which they rest. If they did not, what advantage would camouflaged wings confer? Failure to choose a correct background might lead to death, so Kettlewell believed that background choice must occur. Although skeptical of laboratory experiments, Kettlewell designed a simple test of his hypothesis. He lined a large cider barrel with overlapping strips of black and white cloth. In the evening he released equal numbers of dark- and light-winged moths in the barrel. The top of the barrel was then covered with a sheet of glass and a white cloth. In the morning, the resting position of each moth was recorded. Kettlewell obtained the following results:

	Dark-winged Moths	Light-winged Moths
Black Background	38	20
White Background	21	39

Analyzing the results statistically with a x^2 test, Kettlewell found that the differences in background choice were highly significant. He then attempted to further test his hypothesis by observing background choices made by moths in the field. To his disappointment, the field studies showed no statistically significant tendency for moths to choose correct backgrounds. Moths seemed to land randomly on correct and incorrect backgrounds. Despite these negative results, Kettlewell continued to believe in the importance of background choice.

PROBLEM
In what ways would the natural environment of a moth be different from Kettlewell's laboratory experiment? How might these factors explain the different results obtained in the two experiments? Was Kettlewell justified in holding his background choice hypothesis even though some of his experimental results were negative?

Background choice turned out to be a dead end in Kettlewell's research. Although he continued to believe in it, he could never convincingly demonstrate that it occurred. More important, as it turned out, was his claim that birds selectively prey upon conspicuous moths. In another early experiment Kettlewell released equal numbers of dark- and light-winged moths in a large, outdoor aviary containing light and dark tree trunks. After the moths had come to rest on the tree trunks, Kettlewell released a pair of insectivorous birds. During the first two hours no moths were eaten, even those resting on contrasting backgrounds. But once the birds learned to recognize the moths, they actively searched for conspicuous prey. Inconspicuous moths were less often eaten, although they, too, were sometimes killed, particularly if they happened to be resting near a conspicuous individual.

From this experiment, Kettlewell concluded that birds could act as selective agents, but predation was a learned behavior. The birds had to learn to recognize a specific type of food before they could effectively exploit it. Other biologists remained skeptical. Both ornithologists and entomologists denied that birds would actively search for moths on the basis of wing coloration. In response to this criticism Kettlewell enlisted the aid of the ethologist Niko Tinbergen (see Chapter 14), who took still photographs and motion picture films of predatory birds in the field. His observations revealed that birds captured conspicuous moths approximately three times more often than inconspicuous moths resting on the same tree trunk. This provided dramatic evidence to support Kettlewell's hypothesis.

KETTLEWELL'S FIELD EXPERIMENTS

Kettlewell provided even stronger evidence for natural selection with a series of mark-release-recapture experiments conducted in two different environments: a polluted forest near the industrial city of Birmingham and a pristine forest in rural Dorset. Following methods pioneered by Fisher and Ford, Kettlewell marked the undersides of the wings of male moths with dots of paint. Only males were used because females rarely fly and are, therefore, difficult to recapture. After marking, large numbers of light- and dark-winged moths were released at sundown. Every evening for the next week, males were recaptured using mercury vapor lamps and pheromone traps (i.e., traps containing virgin females). The following table summarizes the results of some of the experiments conducted in the two contrasting environments. The number of recaptures is expressed as a fraction of the total number of moths released.

Moths	Environment	
	Polluted Woods (Birmingham)	Unpolluted Woods (Dorset)
Light wings	18/137 (13%)	62/496 (12.5%)
Dark wings	136/493 (27.5%)	34/488 (7%)

This appears to be a classic example of a controlled experiment. Regardless of the environment, the camouflaged moths are approximately twice as likely to be recaptured in traps as the conspicuous moths. Kettlewell concluded that the missing conspicuous moths had been eaten by predatory birds. In retrospect, this clever set of experiments provides convincing evidence that industrial melanism is caused by natural selection. But how did Kettlewell's contemporaries respond to the experiments? The actual history of the case is more revealing than many textbook accounts would lead us to believe.

In 1952 and 1953, Kettlewell conducted the first series of mark-release-recapture experiments in the polluted forest near Birmingham. The comparable experiments in an unpolluted forest in rural Dorset were done several months later, and the results of the two sets of experiments were published separately. Readers of Kettlewell's first article encountered data from the Birmingham experiment, but they had no way of knowing about the results from the contrasting environment.

PROBLEM
Suppose that you have just read the report of Kettlewell's mark-release-recapture experiments near Birmingham. The results from Dorset have not been published, so you do not know about the second set of experiments. Considering only the Birmingham data, what alternative hypotheses (other than predation) might explain why dark-winged moths are captured more frequently than light-winged moths? How do the combined data from two environments make these alternative explanations unlikely?

There are several plausible explanations for Kettlewell's decision to publish the two sets of data separately. Finding comparable woodlands in polluted and unpolluted areas of England was difficult, and at first he may not have thought it was necessary to duplicate the experiment in different environments. Conducting large field experiments is laborious, and Kettlewell usually had little help (his wife and son were often his only assistants). Traveling between two study sites, both many miles from Oxford, was time consuming. Breeding hundreds of moths to be released at the same time also posed practical problems. All of these factors prevented Kettlewell from conducting the complete set of experiments at the same time.

Whatever the reasons, Kettlewell's early paper gives little indication that it was written as a preliminary report. Nowhere in the paper did Kettlewell discuss the need for a parallel experiment in an unpolluted environment. It seems likely that he initially believed that the experiment in a polluted woods was sufficiently compelling to support his hypothesis. Perhaps he thought that every alternative explanation for his results had been effectively refuted. For example, he found no evidence to suggest that dark-winged moths were more likely to enter traps than light-winged moths. Nor did he find a greater tendency for light-winged moths to migrate out of the study area. Both types of moths seemed equally hearty; light-winged moths were no more likely than their dark-winged counterparts to die from causes other than predation. By doing these checks, Kettlewell believed that he had eliminated all possible variables other than the one he was testing.

Apparently other biologists found the initial experiment unconvincing. Kettlewell later recalled that the results of his Birmingham experiment encountered considerable skepticism from his contemporaries. Therefore, he felt compelled to repeat the experiment in a contrasting environment. The combined results, together with Tinbergen's films of birds selectively eating conspicuous moths, convinced most biologists that Kettlewell's explanation was correct. Natural selection, the result of selective predation by birds, was the most likely cause of industrial melanism.

RECONSIDERING KETTLEWELL'S EXPERIMENTS

Kettlewell's experiments are important for several reasons. It is often incorrectly assumed that evolutionary hypotheses cannot be tested by experiments. Kettlewell's research demonstrates how false this belief is. His large-scale field experiments were done with the degree of care and precision usually associated with laboratory science. As a result, we now know that intense natural selection can sometimes lead to rapid evolution in populations.

Like other great experiments, Kettlewell's work raised as many questions as it answered. If his explanation for industrial melanism was correct, what would happen if pollution was reversed? If smoke and soot were eliminated, would the selective advantage shift away from melanic moths and once more favor light-winged individuals? Biologists who have studied populations of peppered moths in industrial areas where pollution has been reduced find that light-winged individuals are once again on the increase.

Finally, in reconsidering Kettlewell's approach to doing biology it is important to remember that he used a variety of evidence to support his hypothesis. When he started, there were a number of possible explanations for industrial melanism. Natural selection through predation was a likely possibility, but it needed to be conclusively demonstrated. During the course of several years, Kettlewell combined several techniques to find supporting evidence for natural selection. Some of his approaches turned out to be dead ends and were abandoned. Others, like Tinbergen's movies, were dramatic, but only when combined with experimental evidence. The mark-release-recapture experiments began as a small part of Kettlewell's project but grew into the most important part. The combined experiments in two different environments turned out to be the key for solving the problem of industrial melanism.

☐ *EPILOGUE*

Kettlewell often referred to industrial melanism as Darwin's "missing evidence" for natural selection. Could this same process lead to speciation, as Darwin claimed, and could speciation occur rapidly enough for scientists to witness the origin of new species? In the case of the peppered moths this has not happened. There is no evidence that reproductive isolation has evolved in the moth populations. Dark- and light-winged individuals interbreed freely. Some evolutionary biologists believe that even with intense natural selection, speciation could not occur unless a geographical barrier prevented populations of light- and dark-winged moths from mating. Other evolutionary biologists question the need for such barriers and believe that speciation can sometimes occur even without geographical isolation. Perhaps it can happen quite rapidly.

One of the most intriguing candidates for such rapid speciation is the fruit fly, *Rhagoletis pomonella*, a common agricultural pest. Females lay eggs on apples and other related fruit, and the maggots ruin the fruit by feeding on it. Originally, the hosts for this parasite were hawthorns, small trees widely distributed throughout the eastern United States. Thanks to John Chapman ("Johnny Appleseed") and other pioneers, extensive orchards were planted throughout Ohio, Indiana, and Illinois during the early decades of the nineteenth century. *R. pomonella* rapidly colonized the new hosts: apple, cherry, and pear trees.

When evolutionary biologists began studying *R. pomonella* during the 1970s, they discovered that each host species seemed to harbor a genetically distinct population of the parasitic flies. These differences appeared to be maintained partly because fruit flies prefer the type of tree on which they hatch. Females raised on apple trees usually lay their eggs on apples, and females raised on hawthorns usually lay their eggs on the fruit of hawthorns. Males search for mates on the host where they hatch. Furthermore, maggots develop at different rates on the two hosts. Maggots living on apples develop in about 40 days, but those living on hawthorns take 55 to 60 days to develop. As a result, fruit flies on the two hosts become sexually mature at different times.

Despite these important behavioral and physiological differences, reproductive isolation is not complete in populations of *R. pomonella*. When flies from different hosts are brought together in the laboratory, they freely interbreed. Yet the populations have diverged in some important reproductive characteristics during a remarkably short period of time (about 100 fruit fly generations). Perhaps this is a case of speciation in action. Alternatively, reproductive isolation may never evolve completely, and despite their partial isolation the different populations may continue to form a single species. The problem continues to challenge evolutionary biologists.

Another unsolved problem in the case of *R. pomonella* is whether natural selection has altered populations. Apples and hawthorns provide two different habitats for fruit flies, and it stands to reason that selection might act differently in the two environments. But the genetic differences may also be due to chance. Discovering the cause of evolutionary change in fruit flies is another ongoing research problem for evolutionary biologists.

QUESTIONS AND ACTIVITIES

1. What does this case show about the following aspects of doing biology?
   — alternative interpretations of experimental data
   — difficulty of designing controlled field experiments
   — importance of experimental controls
   — different types of evidence used to support theories

2. Reconsider the early experiments on industrial melanism conducted by J. W. Heslop Harrison. If Harrison's hypothesis had been correct, what would be the rate of mutation in his experimental populations of moths? R. A. Fisher pointed out that most naturally occurring mutations appear in approximately 1 in 10,000 individuals. How might Harrison have responded to Fisher's criticism?

3. Kettlewell always believed that background choice was an important factor in the evolution of industrial melanism. How would background choice affect natural selection in peppered moths? Is background choice really necessary for natural selection to occur?

4. Kettlewell and other biologists found that industrial melanism is common in moths, but not in butterflies. What behavioral differences between these two related groups of insects might explain this observation?

5. Consider the data from Kettlewell's mark-release-recapture experiments in Birmingham and Dorset. In both experiments, camouflaged moths were about twice as likely to be recaptured as conspicuous moths. But the relative numbers (%) of recaptured moths were quite different in the two contrasting environments. How can you explain these overall differences in recapture rates?

SUGGESTED READING

Bishop, J. A., and D. P. Clark. 1980. "Industrial Melanism and the Urban Environment." *Advances in Evolutionary Studies* 11: 373–404.

Ford, E. B. 1980. "Some Recollections Pertaining to the Evolutionary Synthesis." In E. Mayr and W. B. Provine, eds., *The Evolutionary Synthesis: Perspectives on the Unification of Biology.* Cambridge, MA: Harvard University Press.

Kettlewell, H. B. D. 1973. *The Evolution of Melanism: The Study of a Recurring Necessity.* Oxford, England: Oxford University Press.

Provine, W. B. 1971. *The Origins of Theoretical Population Genetics.* Chicago: University of Chicago Press.

Ridley, M. 1993. *Evolution.* Boston: Blackwell.

Smocovitis, V. B. 1992. "Unifying Biology: The Evolutionary Synthesis and Evolutionary Biology." *Journal of the History of Biology* 25: 1–65.

Weiner, J. 1994. *The Beak of the Finch: A Story of Evolution in Our Time.* New York: Knopf.

CHAPTER 2

Robert Whittaker & the Classification of Kingdoms

JOEL B. HAGEN

☐ INTRODUCTION

Imagine walking through a tropical rain forest in Central America. Above your head, a howler monkey roars in disapproval as it hangs from the limb of a *Cecropia* tree. A toucan with its outlandish yellow beak flies lazily above, visible through a small break in the forest canopy. With a brilliant flash of metallic blue an emperor butterfly flits in front of you. A column of leaf-cutter ants crosses the trail, each worker carrying a triangular piece of leaf like a parasol above its head. Unseen beneath your feet, billions of soil bacteria and fungi rapidly digest any leaf that falls from the canopy above, recycling the nutrients to the trees. These are just a few examples of the more than two million species of organisms recognized by biologists. Some taxonomists believe that perhaps thirty million more remain to be discovered. How are these multitudes of living things related? How do we create order out of this chaos of diversity? These are two of the most fundamental questions that biologists have always faced.

Traditionally, biologists divided all living things into two large groups: the plant and animal kingdoms. Many nonbiologists still see the living world this way. Yet, during the Renaissance, when the boundary between the living and nonliving worlds was less sharply drawn, it was common to place minerals into a third kingdom equivalent to the plant and animal kingdoms. Throughout history other three- and four-kingdom systems have occasionally been suggested. Today most biologists favor a five-kingdom system proposed about 25 years ago by the ecologist Robert H. Whittaker.

Despite its current popularity, the logic of this new system was not immediately obvious even to Whittaker himself when he began studying taxonomy in 1957. It took him over ten years to work out the details, and he continued to tinker with the system until his death in 1980. Why did our ideas about classification change so dramatically in response to Whittaker's research during the late 1950s and 1960s? Did the change to a five-kingdom system simply reflect our growing knowledge about the living world around us? If so, is the current system the correct way to classify organisms? Or, as some critics of taxonomy have charged, is classification merely glorified "stamp collecting," where the choice between alternative systems is simply a matter of taste?

Maybe there is another way to look at taxonomy. Perhaps classification systems are like other scientific theories that often change over time. If so, we might expect that at different points in history, biologists would favor different systems of classification. Their choices would partly reflect the state of knowledge about the living world but also reflect the current interests of biologists. As new areas of research emerged, ideas about classification would also change in response. This view of taxonomy as a creative, problem-solving activity is well illustrated by the development of Whittaker's five-kingdom system.

ROBERT WHITTAKER: AN EMINENT ECOLOGIST

When he died of cancer in 1980, Robert Whittaker was one of the most influential ecologists of his day. Aside from his development of the five-kingdom system of classification, he wrote nearly 150 books and articles on almost every important topic in plant ecology. In 1974 he was elected to the National Academy of Sciences, one of the greatest honors that an American scientist can achieve. Shortly before his death, Whittaker was named "Eminent Ecologist" by the Ecological Society of America, the highest recognition of success awarded by that professional association.

Despite his success, Whittaker's career started inauspiciously. Unimpressed by his undergraduate transcript, the Botany Department at the University of Illinois rejected Whittaker's application to graduate school. He was later admitted to the Zoology Department, where he completed a Ph.D. dissertation in community ecology. This study of the distribution of plant communities in the Smoky Mountains of Tennessee eventually became a classic paper in ecology. It is perhaps ironic that such an important study of plant communities (animals are not mentioned in the paper) was completed in a zoology department by a student who could not gain admittance to the graduate program in botany. This episode is instructive for two reasons. First, it shows that sometimes it may be difficult for teachers to recognize scientific potential in a student. Second, it highlights the artificiality of the boundary between zoology and botany. Like many other ecologists of his generation, Whittaker was interested in broad biological problems that did not fit the traditional distinction between plants and animals. His later work on the five-kingdom system of classification is a particularly good example of this broader view of the living world.

Every successful scientist has a distinctive approach to research. Those who knew Whittaker remember the intensity of his personality. He could immerse himself in a new area of research, master the literature of the subject, and create a novel theoretical explanation. The breadth of his biological interests was quite unusual, because most scientists focus on rather narrow lines of research. Many of Whittaker's most important papers involved blunt criticisms of well-accepted ideas and competing theories. This tactic could be intimidating, and some scientists accused him of being arrogant, dogmatic, and overly aggressive in his analyses of opposing views. Early in his career, his scientific style prevented him from publishing some of his research—it took eight years to convince an editor to publish his dissertation. It may also have cost him his first teaching position (he apparently criticized senior professors in his department—a dangerous move for an untenured instructor).

Personality can also influence a scientist's work. Whittaker had a reputation as a maverick in ecological theory, and he often had to defend unpopular ideas. His interest in broad biological problems meant that he sometimes moved out of his own specialty and argued with experts in other fields. Such a strategy involves professional risks, but Whittaker never shied from controversy. He actively pursued very broad, interdisciplinary problems. Perhaps if he had been less combative, his five-kingdom system would not have been successful.

A PRELIMINARY TAXONOMIC SCHEME

When Whittaker began his work, the reigning taxonomic system divided all organisms into the plant and animal kingdoms. Like other reformers before him, Whittaker criticized this system because it did not accurately reflect important biological relationships. Fungi, bacteria, and other distantly related organisms were lumped together in the plant kingdom (Figure 2.1). Animals, for the most part, were easily characterized, but what was to be done with curious creatures such as *Euglena viridis*, which shared characteristics of both plants and animals? Traditionally, both botanists and zoologists had claimed these unusual unicellular organisms that photosynthesize and also ingest food.

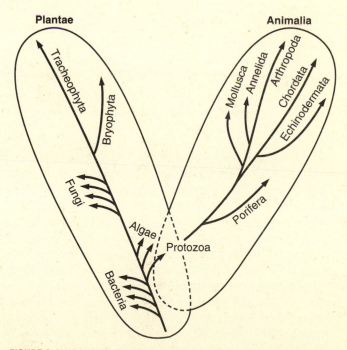

FIGURE 2.1 Whittaker's pictorial representation of the traditional two-kingdom system. Only some of the major phyla are included in this diagram.

Whittaker was equally critical of some earlier attempts to expand the number of kingdoms, particularly a four-kingdom system proposed by H. F. Copeland in 1956. Copeland recognized a new kingdom Monera (or Mychota) for all bacteria and a kingdom Protista (or Protoctista), which included various algae, protozoans, and fungi.

Copeland tried to represent as accurately as possible the important phylogenetic (evolutionary) relationships within these four fundamental groups of organisms. Each kingdom formed a major branch on the evolutionary tree, all members of which were descended from a common ancestor. In other words, all of the members of each kingdom were more closely related to one another than to any members of other kingdoms. The technical term for such a group is **monophyletic**. In a **polyphyletic** kingdom, by contrast, some organisms would be more closely related to some members of other kingdoms than to some members of their own (Figures 2.2(A) and 2.2(B)). By analogy, if you grouped two cousins together because they both have blue eyes but, in the process, separated a brother and a sister because one has blue eyes and one has brown, you would have created something similar to polyphyletic groups. Most taxonomists insist on monophyletic groups because they accurately reflect evolutionary relationships.

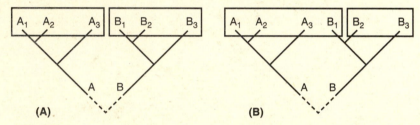

(A)   (B)

FIGURE 2.2 (A) An evolutionary tree with two branches leading to six present-day groups. These groups can be classified into two larger groups (A_1, A_2, and A_3) and (B_1, B_2, and B_3). These larger groups are monophyletic because all members of the first group descended from A, and all members of the second group descended from B. (B) In this case the group (A_1, A_2, A_3, and B_1) is polyphyletic because the members are not all descended from A. From an evolutionary perspective, this classification is invalid even though A_3 and B_1 may be superficially quite similar.

Although Copeland's proposed system followed well-accepted principles of taxonomy, Whittaker argued that it flew in the face of equally well-established ecological principles. In a short two-page note published in the journal *Ecology*, Whittaker pointed out that ecologists already had a *functional classification system* based upon the roles that organisms play in an ecosystem. Although the groups overlap a bit, organisms can generally be classified as producers, consumers, and decomposers. **Producers** use sunlight to synthesize carbohydrates. **Consumers** obtain food by eating living organisms. **Decomposers** feed on dead organic material, breaking macromolecules down into small, inorganic compounds. In a general way, these ecological groups correspond to major taxonomic groups of multicellular organisms: plants are producers, animals are consumers, and fungi are decomposers. Thus, according to Whittaker, ecological function provided a coherent basis for classifying most organisms that biologists study (Figure 2.3).

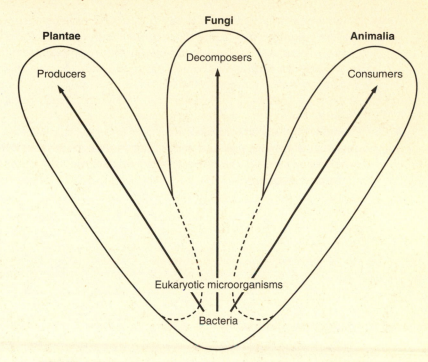

FIGURE 2.3 Whittaker's early three-kingdom system based upon ecological function.

Whittaker believed that this ecological classification system also reflected three major branches on the evolutionary tree. Each branch has evolved in response to a different type of nutrition. Leaves, roots, and conducting tissue in plants have evolved to optimize photosynthesis. Animals, as ecological consumers, have evolved various adaptations for catching and ingesting prey. Fungi, which often live on decaying organic material in the soil, have evolved bodies of ramifying filaments embedded directly in the food supply. Unlike animals, fungi absorb nutrients that they have digested externally.

The ecological and evolutionary justifications for this tripartite division were appealing, but Whittaker realized that the system also had some serious problems. In order to classify all organisms as producers, consumers, and decomposers, Whittaker would be forced to place most bacteria in kingdom Fungi. This made sense ecologically, because most bacteria are decomposers and like the fungi absorb nutrients from a food source that is externally digested. Combining the fungi and bacteria could not be justified on evolutionary grounds, however, because as Whittaker acknowledged, bacteria are no more closely related to fungi than to plants or animals.

Another problem involved more complex unicellular organisms such as algae and protozoans. These creatures are ecologically and phylogenetically diverse. Copeland had placed most unicellular organisms into kingdom Protista, but this kingdom also contained fungi and some other large, multicellular organisms. In his original paper, Whittaker rejected this new kingdom because it would unite ecologically diverse organisms. From a functional point of view, Whittaker believed, all microorganisms acted either as plants, animals, or fungi. He was confident that most unicellular organisms could find a home in one of his three ecologically defined kingdoms.

PROBLEM

Consider the similarities and differences among the organisms in the table. What are the problems with placing euglenoids and fungi (for example, mushrooms) in the plant kingdom? How well did Copeland's four-kingdom system and Whittaker's initial three-kingdom system solve these problems?

	Flowering Plant	Euglenoid*	Mushroom	Vertebrate Animal
Cell wall present	Yes	No	Yes	No
Cell wall material	Cellulose	—	Chitin	—
Cells have flagella	No	Yes	No	Some
Ingest food (heterotrophic)	No	Some	No	Yes
Absorb food (heterotrophic)	No	Some	Yes	No
Photosynthetic (autotrophic)	Yes	Some	No	No
Multicellular organisms	Yes	No	Yes	Yes
Sexual reproduction	Yes	No	Yes	Yes
Energy stored as starch	Yes	No	No	No

*Approximately 1,000 species, including the familiar green flagellate, *Euglena viridis*.

REFINING THE SYSTEM: AN ALTERNATIVE FOUR-KINGDOM PLAN

Whittaker had the germ of an important idea in 1957, but he had worked out few of the details. After his first article, Whittaker immersed himself in the taxonomic literature, particularly the classification of unicellular organisms, which he knew little about. Two years later he admitted that these organisms could not simply be distributed among the kingdoms Plantae, Animalia, and Fungi. A new kingdom Protista would need to be created, but it would be defined quite differently than Copeland's kingdom of the same name. According to Whittaker, all protists must be unicellular. He divided this new kingdom into two parts. The higher protists included all nucleated (**eukaryotic**) unicellular organisms: protozoans, diatoms, euglenoids, and many other microscopic organisms. Nonnucleated (**prokaryotic**) cells made up a lower subkingdom (Figure 2.4). This group included both the true bacteria and the cyanobacteria, an important group of photosynthetic prokaryotes often, but incorrectly, referred to as "blue-green algae." Unlike Copeland, Whittaker excluded all of the fungi, marine algae, and other multicellular organisms from kingdom Protista.

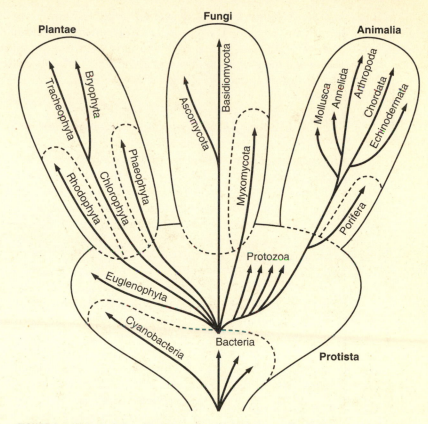

FIGURE 2.4 Whittaker's four-kingdom system. In this scheme, all unicellular organisms belong to kingdom Protista, which is divided into higher and lower subkingdoms.

Whittaker justified his new kingdom because he believed that unicellular organisms formed a distinct evolutionary level or grade. During the distant past, unicellular organisms filled all three fundamental ecological roles: producers, consumers, and decomposers. Some of the unicellular lines had evolved into complex, multicellular organisms, so today most ecosystems are dominated by plants, animals, and fungi. But modern protists—the direct descendants of early unicellular organisms—still carry out these ecological roles in some ecosystems.

Drawing the boundaries of kingdom Protista was troublesome. Whittaker admitted that bacteria were structurally much simpler than the higher protists. Why not place the two groups in separate kingdoms? Such a decision would undermine the ecological basis for defining his kingdoms. Perhaps also for simplicity's sake, Whittaker declined to recognize a separate kingdom Monera—at least in 1959. The upper boundaries of kingdom Protista were also quite fuzzy. Many unicellular protists were very closely related to multicellular plants, animals, or fungi. According to

Whittaker, these boundary problems were inevitable. During the course of evolution some unicellular lineages remained relatively unchanged, while others split into closely related unicellular and multicellular groups. Knowing this, however, did not necessarily make the job of classification any easier. Throughout the 1960s, Whittaker wrestled with the problems of classifying the protists.

COMPLETING THE FIVE-KINGDOM SYSTEM

During the years when Whittaker was working out these problems in his classification system, he was at the height of his professional career. He wrote a string of influential research articles on several critical ecological topics. He completed one of the most popular textbooks of ecology, which was widely read by both students and professional biologists. At the end of the decade (1969) he also presented the culmination of his taxonomic research in *Science*, perhaps the most prestigious scientific journal in the world.

During the decade following his 1959 paper on classification, Whittaker had become convinced that unicellular organisms must be divided into two separate kingdoms. He placed prokaryotic cells, which lack a true nucleus and most specialized organelles, into kingdom Monera. Eukaryotic, unicellular organisms, each with a nucleus and many specialized organelles, remained in kingdom Protista.

This important distinction between two basic types of cells was the last innovation that Whittaker added to his system, although he had hinted at the possibility in earlier papers. His thinking had been heavily influenced by Lynn Margulis's theory of endosymbiosis (see Chapter 3). Contrary to the prevailing view that protists had gradually evolved from prokaryotic bacteria, Margulis claimed that all eukaryotic cells were derived from multiple symbiotic partnerships among prokaryotic cells. According to this theory, some of the specialized organelles of eukaryotic cells had originated as free-living bacteria. If this explanation of cellular evolution was true, then the transition from prokaryotic cells to eukaryotic cells must have occurred relatively quickly. There should be few intermediate forms between the types of cells, and, therefore, the boundary between kingdoms Monera and Protista should be sharply defined (Figure 2.5).

Endosymbiosis was only a provisional theory in 1969, but it certainly strengthened Whittaker's five-kingdom system. All organisms could now be arranged hierarchically into three well-defined evolutionary levels: prokaryotic organisms (kingdom Monera); eukaryotic, unicellular organisms (kingdom Protista); and eukaryotic, multicellular organisms (kingdoms Plantae, Animalia, and Fungi). Upon this evolutionary hierarchy, Whittaker superimposed his original ecological classification based on nutrition: producers, consumers, and decomposers. These ecological distinctions can be seen in the horizontal arrangement of the multicellular kingdoms and in the various evolutionary lines within kingdom Protista (Figure 2.5).

By 1969 the broad outlines of a successful taxonomic system emerged. Because Whittaker used a variety of criteria—evolutionary, ecological, cellular, and molecular—his system appealed to a broad audience of biologists. Compared to the traditional two-kingdom system and Copeland's four-kingdom system, Whittaker's five

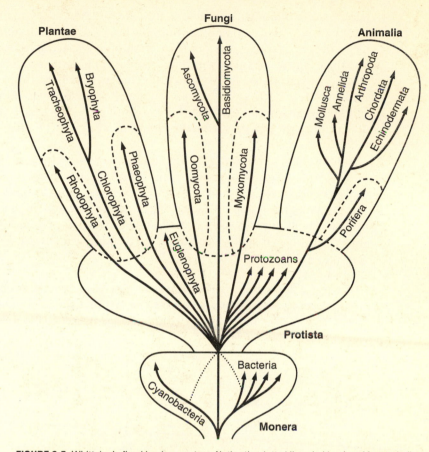

FIGURE 2.5 Whittaker's five-kingdom system. Notice the dotted lines in kingdom Monera indicating the endosymbiotic origin of eukaryotic cells. Also notice that the boundary between kingdoms Monera and Protista is very narrow, because according to the endosymbiotic theory there are few intermediaries between prokaryotic and eukaryotic cells. Ambiguous "problem groups" included in plants, fungi, and animals make each of these multicellular kingdoms polyphyletic in Whittaker's scheme.

kingdoms seemed to be based upon fundamental natural relationships. Whittaker emphasized the differences between his system and its two competitors. By aggressively presenting his argument in a prestigious scientific journal, Whittaker ensured that his critical comparison of taxonomic systems reached a broad audience.

In contrast, neither of the other competing systems had vocal supporters. Few biologists in the later 1960s would strongly defend the outmoded two-kingdom system. Copeland never responded to Whittaker's early criticisms of his ideas, and he died in 1968, a year before Whittaker unveiled the final version of the five-kingdom system. Many biologists initially sympathetic to the four-kingdom system switched allegiance to Whittaker's plan. Within a few years, therefore, the five-kingdom system was almost universally adopted by biologists.

It might be tempting to conclude that in 1969 Whittaker finally discovered the correct way to classify organisms. But he was quick to admit that important problems remained with his system. For example, what was to be done with the green algae (Chlorophyta), an important group of aquatic producers whose ancestors also gave rise to all land plants? The green algae (approximately 7,000 species) include both unicellular and multicellular forms. Do they belong in kingdom Plantae or kingdom Protista? Because they are so closely related to higher plants, Whittaker placed the green algae in kingdom Plantae—a decision that remains controversial.

Equally controversial was his decision to place the red algae (Rhodophyta) and brown algae (Phaeophyta) in the plant kingdom. He reasoned that even though these large, multicellular seaweeds are not closely related to other plants, they play the same ecological role—they are "functional plants" in many marine ecosystems. Furthermore, because of their size and complexity, they do not fit the unicellular characteristic that Whittaker used to define his kingdom Protista.

PROBLEM

Consider the characteristics of the groups in the table. What characteristics could be used to place all three groups into the plant kingdom? On what basis could you exclude the green algae from the plant kingdom? On what basis could you exclude the brown algae?

	Flowering Plants	Green Algae	Brown Algae
Cellulose in cell wall	Yes	Yes	Yes
Forms of chlorophyll	a and b	a and b	a and c
Energy stored in starch	Yes	Yes	No
Vascular tissue	Yes	No	Yes
Multicellular	Yes	Some	Yes
Cells with flagella	No	Some	Reproductive cells
Habitat	Mostly terrestrial	Mostly freshwater	Mostly marine

Whittaker realized that a plant kingdom including the seaweeds would be polyphyletic. The shared similarity in ecological roles between land plants and seaweeds seemed too compelling to ignore, however. As an ecologist, Whittaker was willing to accept a polyphyletic plant kingdom, but most taxonomists found this unacceptable. They modified Whittaker's system by placing the red and brown algae into kingdom Protista—a decision that makes this kingdom a hodgepodge of unicellular and multicellular organisms.

Whittaker was aware of the problems facing his five-kingdom proposal, and he continued to debate the issues until his death in 1980. He struggled with the conflicting demands of a system that would reflect important ecological principles while still accurately portraying evolutionary relationships. Just as important was the demand that the system be convenient to use, both by students and professional biologists. One critic, for example, claimed that a truly monophyletic classification of protists would require 20 separate kingdoms. Such a system might accurately reflect evolutionary history, but Whittaker pointed out that it would be needlessly cumbersome. A five-kingdom system was more reasonable, but it meant compromising some well-established taxonomic principles and accepting many ambiguities.

☐ *EPILOGUE*

According to Whittaker, the uncritical acceptance of the two-kingdom system before 1969 was largely due to a distinction that humans make between the large organisms they most often encounter: plants and animals. This common-sense dichotomy became enshrined in the organizational structure of biology. Most biologists identified themselves either as botanists or zoologists, and many university departments were organized along these lines. The newer five-kingdom system called into question the logic of the plant-animal dichotomy. Although many universities still have botany and zoology departments, most biologists now recognize that this distinction is artificial and does not reflect the most important boundaries in the natural world.

In Whittaker's five-kingdom system a new dichotomy was recognized. The distinction between prokaryotes and eukaryotes became the most important boundary in the living world. All organisms seemed to fit neatly into one of these two categories, and Lynn Margulis's endosymbiotic theory explained why this should be the case. What were the consequences of this change? Some biologists have emphasized the positive influence that Whittaker and Margulis had on directing research toward previously little-studied groups of unicellular organisms. Better understanding the prokaryote-eukaryote boundary became an important biological problem. But other biologists claim that this distinction became a "new dogma" and that the biology of prokaryotes was taken for granted by biologists who were primarily interested in eukaryotic cells and how they evolved.

Carl Woese has spent his career studying several unusual groups of prokaryotes referred to as the archaebacteria. These organisms have fundamentally different molecular, biochemical, and ecological characteristics than other bacteria. The archaebacteria are intriguing because they often exist in extreme habitats characterized by high temperatures, high salinity, and high acidity. Some (methane producers) are of vital human importance. Largely due to the work of Woese and his colleagues, most biologists now recognize the archaebacteria as a separate subkingdom. Woese, however, believes that a more radical restructuring is in order. According to Woese, archaebacteria are so different from all other organisms that they should be placed in a separate taxonomic group. He would erect a new system with three superkingdoms or "domains": Archae (archaebacteria), Bacteria (all other bacteria), and Eucarya (all eukaryotic organisms). Although most biology textbooks ignore this proposal, a few now place the archaebacteria into a sixth kingdom. The ongoing debate over the status of these unusual prokaryotes serves to warn us that systems of classification, like all scientific theories, are open to revision.

QUESTIONS AND ACTIVITIES

1. What does this case show about the following aspects of doing biology?
   — revision of scientific theories
   — role of assumptions in creating new theories
   — interrelationships among scientific disciplines
   — the role of personality in scientific creativity

2. The two-kingdom system was accepted by most biologists and naturalists for hundreds of years. Why do you think this way of classifying the living world was so popular?

3. Why do you think fungi and bacteria were originally classified as plants rather than animals? What changes in biology made this classification unsatisfactory?

4. Many evolutionary biologists reject the use of comparative terms such as *higher* and *lower*, or *primitive* and *advanced*, when describing groups. Such comparisons are value laden and may mislead readers into believing that evolution is progressive, always leading to greater complexity. Does Whittaker's five-kingdom diagram have such evolutionary implications? Could the diagram be drawn in a way that does not imply progressive evolution? Would such a diagram be an improvement over Whittaker's originals? *Note:* many different five-kingdom diagrams exist. Compare Figure 2.5 with several different diagrams that you find in biology textbooks.

5. There continues to be disagreement about where to draw the boundaries dividing the five kingdoms. Compare the classification scheme presented in your textbook with others that you find in the library. Do the authors justify their placement of "problem groups" in one kingdom or another? Discuss the advantages and disadvantages of each classification scheme.

SUGGESTED READING

Hagen, J. B. 1992. *An Entangled Bank. The Origins of Ecosystem Ecology.* New Brunswick, NJ: Rutgers University Press.

Margulis, L., and K. V. Schwartz. 1988. *Five Kingdoms: An Illustrated Guide to the Phyla of Life on Earth.* 2nd ed. New York: Freeman.

Rothchild, L. J. 1989. "Protozoa, Protista, Protoctista: What's in a Name?" *Journal of the History of Biology* 22: 277–305.

Sogin, M. L. 1991. "Early Evolution and the Origin of Eukaryotes." *Current Opinion in Genetics and Development* 1: 457–463.

Westman, W. E., and R. K. Peet. 1985. "Robert H. Whittaker (1920–1980): The Man and His Work." *Vegetatio* 48: 97–122.

Whittaker, R. H. 1969. "New Concepts of Kingdoms of Organisms." *Science* 163: 150–160.

Woese, C. R. 1994. "There Must Be a Prokaryote Somewhere: Microbiology's Search for Itself." *Microbiological Reviews* 58: 1–9.

CHAPTER 3

Lynn Margulis $\mathcal{E}$ the Question of How Cells Evolved

JOEL B. HAGEN

☐ INTRODUCTION

Modern biology inherited two great theories from the nineteenth century: evolutionary theory and cell theory. Surprisingly, these theories, so central to our understanding of the living world, have had a rather uneasy relationship. Until quite recently, most cell biologists ignored evolution, and most evolutionary biologists ignored cells. The exception to this historical generalization was the chromosomes, which both evolutionary biologists and cell biologists studied. But what about the cytoplasm, the contents of the cell outside the nucleus? Could knowing about other cellular structures (organelles) add anything to evolutionary theory? Could evolutionary theory suggest interesting questions about the structure or function of organelles? For most biologists, the answer to these questions was no. The cytoplasm added little to understanding evolutionary theory, and vice versa.

Occasionally, some biologists tried to bridge the theoretical gap, but they usually met with derision. For example, during the 1920s the microbiologist Ivan Wallin made the remarkable claim that mitochondria had originated as free-living bacteria. According to Wallin, the former bacteria and their host cells evolved together to establish an inseparable symbiotic partnership. He even claimed to have removed mitochondria from cells and grown them in isolation.

Wallin's idea was almost universally rejected, and he was often ridiculed for his wild speculations. According to his critics, evolution by symbiosis was as improbable as that other great pseudoscientific idea of the 1920s: continental drift (see Chapter 16). Although intrigued by the possibility that mitochondria evolved from bacteria, America's leading cell biologist, E. B. Wilson, remarked that Wallin's ideas were "too fantastic for present mention in polite biological society."

With the benefit of hindsight it is easy to smile at the comparison between continental drift and endosymbiosis, two great scientific heresies that later revolutionized the way we look at the natural world. The criticisms were, however, justified. Wallin's theory was quite speculative. No one, then or now, has verified his claim that mitochondria can be grown outside of cells.

PROBLEM

Assuming that mitochondria really did evolve from free-living bacteria, why might it be difficult or impossible to experimentally grow them outside of the host cell? How can you explain Wallin's unverified claim that he had isolated and grown mitochondria outside of cells?

Both the structure and the function of mitochondria were mysteries in 1920. The internal anatomy of bacteria was also almost totally unknown. The evidence Wallin needed to support his theory required the electron microscope and other sophisticated laboratory techniques developed only after World War II. As in the case of continental drift, the theory of symbiosis in cellular evolution that was finally accepted during the 1970s was very different from the one suggested by Wallin in the 1920s.

LYNN MARGULIS: A REVOLUTIONARY SCIENTIST

Like the eventual acceptance of continental drift, acceptance of a symbiotic theory of cell evolution has often been hailed as a scientific revolution. The woman most responsible for bringing the idea to scientific respectability is Lynn Margulis (Figure 3.1). A prolific writer and dynamic speaker, Margulis captivates audiences and often irritates more traditional biologists with her unorthodox ideas. A profile in *Science* described her as an unruly provocateur, but as one of the world's leading authorities on cellular evolution, she supports her claims with abundant evidence. Although many biologists continue to disagree with some of her ideas, everyone takes endosymbiosis seriously.

FIGURE 3.1 Lynn Margulis. *Source:* Courtesy of Lynn Margulis and the University of Massachusetts Photo Service.

Margulis entered biology during a particularly exciting period. James Watson and Francis Crick were just discovering the structure of DNA when Margulis was in college. A few years later, when she was a graduate student, two of her professors discovered DNA in chloroplasts. Other scientists reported finding DNA in mitochondria (Figure 3.2). Because these early reports were hotly disputed, searching for DNA outside the nucleus was not the sort of research project that most graduate students would have chosen. Despite warnings, Margulis plunged into the controversial problem for her Ph.D. dissertation. Using radioactively labeled nucleotides, she convincingly demonstrated the presence of DNA in the chloroplasts of *Euglena gracilis*, one of the curious unicellular organisms that shares both plant and animal characteristics.

Margulis wrote her first article on the endosymbiotic theory in 1967, two years after she completed her Ph.D. At the time, she was a single mother without a permanent teaching position. She was also writing her first book on endosymbiosis, which sparked a lively controversy when it was published in 1970. Although it initially brought Margulis notoriety, the controversy over cellular evolution was rather short lived. By the time she published a second book on endosymbiosis in 1981, most biologists accepted important parts of her theory. As a result, Margulis became a scientific celebrity whose success was publicized in both popular and professional magazines.

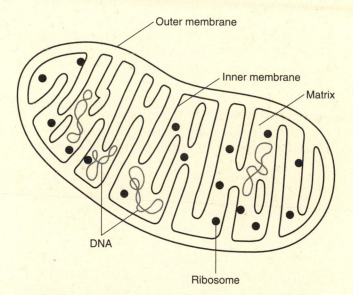

FIGURE 3.2 Cross section of a mitochondrion containing DNA molecules and protein-building ribosomes similar to those found in bacteria. The liquid matrix contains the enzymes responsible for the early steps of respiration (Krebs cycle). Most of the ATP produced during respiration is generated by the oxygen-consuming reactions (electron transport system) that occur on the highly folded inner membrane.

BACKGROUND TO A CONTROVERSY

In 1970, when Margulis's first book was published, most biologists had never heard of endosymbiosis. Those who knew about it usually dismissed it. In order to succeed, Margulis had to carefully distinguish her ideas from the discredited theory proposed by Ivan Wallin half a century earlier. She also had to overcome a basic assumption about evolution held by nearly all biologists at the time. According to the traditional view, evolution usually occurs gradually; endosymbiosis, however, is based on the idea of rather sudden evolutionary changes. Finally, Margulis had to convince biologists to take DNA in the cytoplasm seriously. Although evidence for DNA in chloroplasts and mitochondria was growing stronger, the idea that some genes reside outside the nucleus remained unorthodox.

Despite these biases against endosymbiosis, Margulis's book was widely read. Even those who strongly disagreed with her did not ridicule her theory the way biologists had belittled Ivan Wallin's theory about the evolution of mitochondria. Indeed, the book convinced many biologists that cellular evolution was an exciting, if controversial, field. How had cell biology changed during the 50 years after Wallin proposed his unsuccessful theory?

Much more was known about the internal structure of cells in 1970 than in 1920. Unlike Wallin, who knew little about the internal structure or function of mitochondria, Margulis had access to a great deal of information about the internal structure of cells when she wrote her book. Powerful electron microscopes, perfected after World War II, allowed scientists to study the previously hidden parts of organelles. Using new biochemical techniques, scientists were able to discover many details of cellular activities. Mitochondria, long an enigma, were now known to be important sites of adenosine triphosphate (ATP) production, and for the first time scientists were beginning to understand how this critical process occurred on mitochondrial membranes (see Chapter 8).

By 1970 biologists also became aware of major differences between **prokaryotic** bacteria, which lack nuclei and most other organelles, and **eukaryotic** cells, which have both. The sharp discontinuity between prokaryotes and eukaryotes, which previously had not been fully recognized, was highlighted by Robert Whittaker's new system of classification (see Chapter 2), which used the two cell types to distinguish kingdom Monera from four eukaryotic kingdoms (Animalia, Plantae, Fungi, and Protista). The prokaryotic/eukaryotic distinction was now at the forefront of biological attention. What other similarities and differences might be found between the two types of cells? How had eukaryotic cells evolved? What was the evolutionary significance of the DNA found in some organelles? These were the questions that Margulis set out to answer in 1970.

THE SERIAL ENDOSYMBIOTIC THEORY (SET)

According to Margulis, eukaryotic cells evolved through a series of symbiotic partnerships involving several different kinds of prokaryotic cells. The smaller partners invaded larger host cells and eventually evolved into three different kinds of

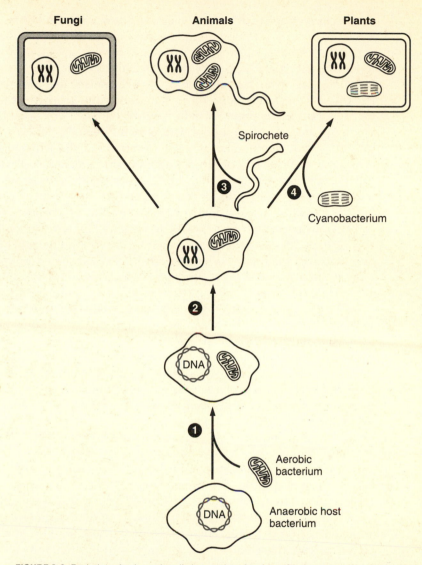

FIGURE 3.3 Evolution of eukaryotic cells by a series of endosymbiotic events: (1) mitochondria evolve from small, free-living, respiring bacteria; (2) the nucleus evolves from the simpler prokaryotic DNA molecule; (3) flagella (undulipodia) evolve from symbiotic spirochetes; (4) chloroplasts arise from free-living cyanobacteria. Cell walls in plants and fungi, which are structurally quite different, evolve independently.

organelles: mitochondria, chloroplasts, and flagella (Figure 3.3). Because these evolutionary steps supposedly occurred as a series of discrete events, Margulis's theory is often referred to as the **SET: serial endosymbiotic theory**.

Like other evolutionary biologists, Margulis believes that life first appeared on the earth about four billion years ago. The first organisms were extremely simple—

microscopic droplets of water containing a few genes and enzymes surrounded by a membrane. They fed on abundant organic molecules that had been produced earlier in the earth's history by various nonliving chemical processes. Like some modern bacteria, early prokaryotic cells extracted energy from these molecules by fermentation, using various forms of metabolism that do not require oxygen. Luckily for the fermenters, there was almost no oxygen in the atmosphere. If there had been, the primitive cells would have been poisoned by this highly reactive gas. Later, as the supply of energy-rich molecules in the watery environment began to be depleted, other types of bacteria evolved which used solar energy to synthesize their own supplies of large, organic molecules. These early photosynthetic bacteria were also anaerobic. In other words, they did not use oxygen and their primitive photosynthetic reactions did not produce oxygen as a by-product. For over a billion years, primitive ecosystems included only two types of prokaryotic organisms: simple photosynthetic bacteria and fermenting bacteria.

Perhaps 2.5 billion years ago, a new group of photosynthetic bacteria evolved, the ancestors of today's cyanobacteria. These advanced photosynthesizers split water to produce the hydrogen ions (H^+) needed to build sugar molecules. A by-product of this water-splitting reaction was oxygen gas. This was a catastrophic event in the history of life. Oxygen is such a reactive element that it easily destroys delicate biological structures. As the amount of oxygen in the atmosphere increased, most species of anaerobic bacteria were driven to extinction, victims of the earth's first case of air pollution. Some survivors retreated to areas of brackish water or other oxygen-depleted habitats, where their anaerobic descendants still flourish today. A few prokaryotes became aerobic by evolving various mechanisms to detoxify oxygen. The most successful of these processes was respiration, which not only converted toxic oxygen back into harmless water molecules, but also generated large quantities of ATP.

According to the SET, the photosynthetic production of oxygen gas and the subsequent evolution of respiration set the stage for the evolution of all eukaryotic cells. This evolutionary process occurred in several separate symbiotic events (Figure 3.3). The first eukaryotic organelles to evolve were mitochondria—structures found in almost all eukaryotic cells. In Margulis's theory, small respiring bacteria parasitized larger, anaerobic prokaryotes. Like some bacteria today (*Bdellovibrio*), these early parasites burrowed through the cell walls of their prey and invaded their cytoplasm. Either the host or the parasite was often killed in the process, but in a few cases the two cells established an uneasy coexistence.

The mutual benefits to the partners are obvious. The respiring parasite, which actually required oxygen, would allow its host to survive in previously uninhabitable, oxygen-rich environments. Perhaps the parasite also shared with its host some of the ATP that it produced using oxygen. In exchange, the host provided sugar or other organic molecules to serve as fuel for aerobic respiration. Eventually, as often occurs with parasites, the protomitochondria lost many metabolic functions provided by the host cell. Similarly, as oxygen in the atmosphere continued to increase, the host became more and more dependent upon its protomitochondria to detoxify the gas. What began as a case of opportunistic parasitism evolved into an obligatory partnership. The small respiratory bacteria eventually evolved into the mitochondria of eukaryotic cells.

Although virtually all eukaryotic cells contain mitochondria, only those of plants and certain protists contain chloroplasts. Therefore, it seems likely that chloroplasts evolved in only a few lines of eukaryotic cells, and this event occurred after mitochondria were already well established. How did this new evolutionary partnership evolve? With higher metabolic rates, cells containing mitochondria were more efficient than anaerobic cells. Some of these newer, unicellular organisms grew larger and evolved into predators capable of eating smaller cells. Their prey undoubtedly included cyanobacteria. In rare cases, these small photosynthetic cells may have resisted digestion after being engulfed. Inside the predator, they set up a semi-independent existence and eventually evolved into chloroplasts.

Although such a scenario may seem far-fetched, we know that similar partnerships exist today. For example, the unusual ciliate *Paramecium bursaria* is host to many unicellular green algae in the genus *Chlorella*. These "pseudochloroplasts" produce sugar molecules that are shared with the host. If the *Chlorella* are experimentally removed, both partners continue to exist independently. Without its photosynthetic partners, however, the *Paramecium* becomes totally dependent upon external sources of food. Provided the opportunity, the *Paramecium* will eat *Chlorella* but will not digest them, thus reestablishing the symbiotic partnership. *Paramecium bursaria* is not a unique case of modern endosymbiosis. Many other organisms, including several multicellular animals, also play host to photosynthetic algae or cyanobacteria.

The most controversial claim made by Margulis is that eukaryotic flagella evolved from small, corkscrew-shaped bacteria called **spirochetes**. Many spirochetes are parasites (the best known, *Treponema pallidum*, causes syphilis). Others are free-living, found in such exotic environments as the intestines of termites. Regardless of how they live, these unusual bacteria swim with an undulating motion reminiscent of the whiplike movement of eukaryotic flagella. Is this similarity evidence for Margulis's evolutionary claim, or is it simply a coincidence? Why not accept the more orthodox explanation that eukaryotic flagella gradually evolved from the simpler flagella found on many bacteria?

Margulis points out that although both types of flagella are used for locomotion, prokaryotic and eukaryotic structures are very different (Figures 3.4(A) and 3.4(B)). Prokaryotic flagella consist of a single, hollow filament of protein that spins on its axis like a tiny propeller. Eukaryotic flagella are much larger; they contain a complex arrangement of 11 microtubules, and the entire structure is surrounded by an extension of the cell membrane. In contrast to the spinning prokaryotic flagellum, the eukaryotic structure propels the cell by lashing back and forth in a whiplike fashion. Because they are so different in structure, function, and perhaps evolutionary origin, Margulis proposes that the eukaryotic flagellum should be referred to by a different term: **undulipodium**.

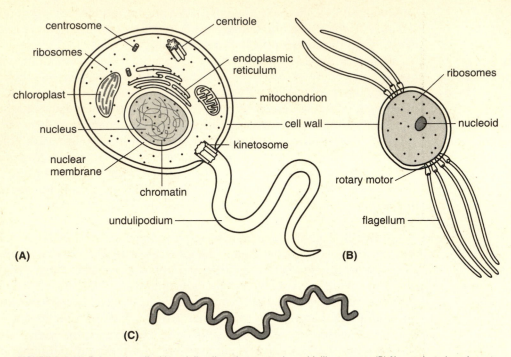

FIGURE 3.4 (A) Eukaryotic cell with undulipodium that moves in a whiplike manner. (B) Nonnucleated, prokaryotic cell with several simple flagella that move by rotation. (C) Spirochete that is propelled in an undulating motion by axial filaments and flagella inside the bacterial cell.

Could undulipodia really have evolved from spirochetes? Margulis claims that the complex arrangement of microtubules in the undulipodium is somewhat similar to long protein filaments (axial filaments) located between the cell wall and an outer sheath membrane in spirochetes (Figure 3.4(C)). Intriguing evidence further supporting her hypothesis comes from cases of "motility symbiosis" described by Margulis. In the hindguts of many termites live a wide variety of protozoans and bacteria, including several types of spirochetes. Biologists had long believed that one of the protozoans (*Mixotricha paradoxa*) was covered with thousands of synchronously beating flagella. Closer examination, however, showed that most of these "flagella" were actually spirochetes regularly arranged in rows on the surface of the protozoan. The rhythmic beating of the symbiotic spirochetes propels the host protozoan through the intestines of the termite.

Mixotricha and other cases of motility symbiosis are intriguing, but the spirochete-undulipodium hypothesis remains far more controversial than the other parts of the SET. Part of the problem is lack of strong supporting evidence. Undulipodia do not contain DNA, RNA, or ribosomes—the remnants of an independent protein-building machinery, which are always found in chloroplasts and mitochondria. If they evolved from free-living bacteria, shouldn't undulipodia also show this evidence of their ancestry? Pointing out this missing evidence, many critics dismiss the spirochete-undulipodium hypothesis as speculation. Margulis claims that critics are too conservative and further research will ultimately confirm her hypothesis. Most biologists remain skeptical, although many admit that some of Margulis's previous "far-fetched ideas" later turned out to be correct.

PROBLEM

Nobel laureate Joshua Lederberg claims that it is impossible to demonstrate convincingly that eukaryotic flagella evolved from spirochetes. Nonetheless, he believes that Margulis's hypothesis is very important. How might an unprovable speculation be useful to scientists?

AUTOGENY: AN ALTERNATIVE THEORY OF CELLULAR EVOLUTION

When Margulis's book appeared in 1970, most biologists accepted the general belief that eukaryotic cells gradually evolved from prokaryotic ancestors. Because cellular evolution was not a major concern of most evolutionary biologists, however, there were few detailed theories to explain how this might have occurred. This situation quickly changed during the 1970s, when Margulis's critics proposed alternative theories of gradual cellular evolution (**autogeny**). Perhaps the best-known of these competing theories was one presented by F. J. R. Taylor, a Canadian botanist.

According to Taylor, all eukaryotic cells evolved through a process of slow, branching evolution. He believed that the original ancestor must have been a photosynthetic bacterium, somewhat similar to the cyanobacteria of today. This must have been so, Taylor reasoned, because cyanobacteria, algae, and plants all use the same form of photosynthesis. It seemed unlikely that this complex process could have independently evolved in each different line. Of course, this means that animals and fungi must have lost the ability to photosynthesize at some later point in evolutionary history (Figure 3.5).

Similar to some cyanobacteria of today, the ancestral cell had a complex system of internal membranes, embedded with the enzymes and pigments used in photosynthesis and respiration, Taylor claimed. Because increasing the surface area of membranes makes a cell more efficient, natural selection favored the evolution of an increasingly elaborate membrane system. Taylor also assumed that DNA was found in several places in the primitive cell. There was a central **nucleoid** where most of the genetic material was located, but several smaller loops of DNA also were scattered throughout the cell. This was a reasonable assumption, because small accessory molecules of DNA are common among bacteria today.

Chloroplasts and mitochondria were formed by two simultaneous evolutionary processes: compartmentalization and specialization. Sections of the elaborately folded membrane system sometimes broke away to form separate, enclosed compartments. Small pieces of DNA and some ribosomes were often trapped inside these bodies. This explains why organelles today contain some protein-making machinery. At the same time, different membrane-bound compartments became specialized to accomplish specific metabolic tasks. Some became centers of photo-

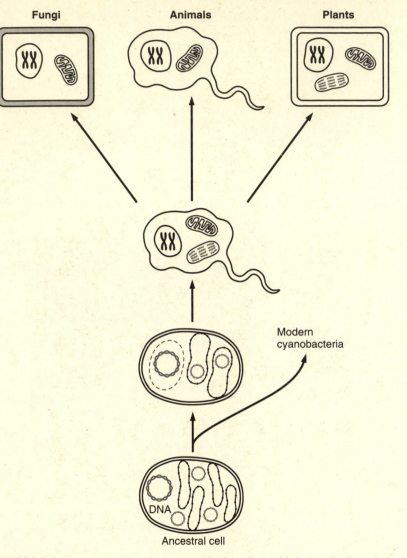

FIGURE 3.5 Autogenous evolution of eukaryotic cells. According to this theory, all cells gradually evolved from a single, primitive photosynthetic cell similar to the cyanobacteria of today. Both mitochondria and chloroplasts evolved from internal membrane systems found in the ancestral cell. Photosynthesis was later lost in the evolutionary lines leading to animals and fungi.

synthesis and eventually evolved into chloroplasts. Others became specialized for respiration and eventually evolved into mitochondria.

During the course of evolution, eukaryotic cells became larger and lost their prokaryotic cell walls. Complex, eukaryotic flagella may also have gradually evolved from simpler bacterial flagella. Presumably, the intermediate forms, which were less efficient for locomotion, disappeared. At least some of the new eukaryotic cells evolved the ability to engulf food particles. These cells became predators, feeding on bacteria and other tiny organisms. The loss of chloroplasts, which occasionally happened, would not necessarily have been a problem for these evolving predators. Other cells evolved new protective outer walls, quite different from prokaryotic cell walls. This adaptation must have evolved quite late and in two separate branches of the evolutionary tree, because the cell walls of plants and fungi are structurally distinct.

RESOLVING THE CONTROVERSY

During the 1970s, scientists were presented with two broad, competing theories of cellular evolution: the SET and autogeny. At the end of this decade, when Margulis published a second book on endosymbiosis, much of her theory was widely accepted. What had been an unpopular theory a decade before was now part of mainstream biology. Autogeny, although not completely dismissed, seemed less plausible. Why was Margulis so successful at persuading her colleagues to radically change their thinking about the evolution of cells? What problems prevented the widespread acceptance of autogeny?

Margulis claims that, unlike Wallin's unsuccessful theory, both the SET and autogeny make predictions that can be tested. According to her, the SET survived these tests while autogeny did not. This is a plausible explanation, but choosing between two general theories involves more than a single experiment or set of observations. Many of the contrasting predictions made by the SET and autogeny are extremely difficult to test. For example, according to the SET, few intermediates between prokaryotic and eukaryotic cells should be found. If, on the other hand, eukaryotic cells evolved gradually, then many "missing links" should have existed. As it turns out, the dividing line between prokaryotes and eukaryotes is quite sharp, which seems to support the SET. But this is not conclusive evidence against autogeny. Although many fossils of unicellular organisms have been found, including those of some of the earliest prokaryotes, there is general agreement that fossilization is a relatively rare event. Therefore, intermediates between prokaryotes and eukaryotes may yet be discovered.

If simple tests of the SET and autogeny provide ambiguous results, how can scientists choose between the two alternatives? Like most complex theories, the SET and autogeny were evaluated not on the basis of a single test but rather on the basis of multiple lines of evidence. If many seemingly unrelated pieces of evidence can be easily explained by one theory but not by the other, scientists will usually choose the stronger theory—even when it cannot be conclusively proven. For example, if the SET is correct, then numerous similarities should exist between free-living bacteria and eukaryotic organelles. Indeed, in some important ways, organelles should be more similar to free-living bacteria than to the rest of the eukaryotic cell. Few, if any, such similarities should be found if eukaryotic organelles evolved autogenously. The success of the SET was largely due to Margulis's ability to demonstrate so many of the expected similarities. Supporters of autogeny had difficulty explaining why these similarities should be found.

As Wallin had pointed out half a century earlier, mitochondria and chloroplasts reproduce by dividing, much the way bacteria do. What he could not know in the

1920s was that the DNA in mitochondria and chloroplasts is structurally similar to the single circular molecule found in bacteria. Unlike DNA in the nucleus, neither bacterial DNA nor DNA in eukaryotic organelles forms chromosomes. The protein-synthesizing ribosomes found in mitochondria and chloroplasts are smaller than those found elsewhere in eukaryotic cells, but they are about the same size as ribosomes found in free-living bacteria. The nucleotide sequences of RNA molecules in the ribosomes of chloroplasts are also more similar to bacterial sequences than to the sequences of other eukaryotic ribosomes. Some antibiotics that interfere with protein synthesis in bacteria have the same effects on mitochondria and chloroplasts but do not inhibit protein synthesis in the rest of the eukaryotic cell. Finally, as would be expected if these organelles evolved symbiotically, cells experimentally deprived of their mitochondria and chloroplasts are unable to reproduce them. Although most of the genetic instructions for making organelles resides in the nucleus, some necessary genes remain in the mitochondria and chloroplasts. None of these individual bits of evidence is convincing. Taken together, however, they make a strong case for some important parts of the SET.

General theories are usually not completely accepted, at least initially. Margulis's SET is an excellent example of this type of piecemeal acceptance. Nearly all biologists, including many of Margulis's original critics, now believe that mitochondria and chloroplasts evolved from bacteria. But most biologists still reject the claim that eukaryotic flagella (undulipodia) evolved from symbiotic spirochetes. It is generally believed that this hypothesis is too speculative and that there is insufficient evidence to support it. Margulis herself admits that this is the most difficult part of the theory to accept, although she believes that eventually enough evidence will accumulate to convince even skeptical biologists.

Whatever the eventual fate of the spirochete-undulipodium hypothesis, the SET has generated much research on cellular structure and function. Through the process of testing and evaluating competing theories, biologists have learned a great deal about cells and how they evolved. Cell biology and evolutionary biology, two fields that previously seemed to have little in common, now share an important, unifying theory.

☐ *EPILOGUE*

A cornerstone of the SET has been Margulis's belief that cellular evolution was primarily the result of adaptations to an aerobic environment. This claim has recently been challenged by some biologists who study rare eukaryotic cells that do not have mitochondria. More than 1,000 species of protozoans and a few species of fungi lack these respiratory organelles.

Some of the protozoans may be "living fossils" similar to the earliest eukaryotic cells. These simple protists also lack such characteristic eukaryotic organelles as Golgi bodies and the endoplasmic reticulum. According to some biologists, similar protozoans may have existed for millions of years before the evolution of mitochondria. Because these unicellular organisms would have been predators, this explains how the bacterial precursors of mitochondria entered their hosts—they were eaten.

But it undermines Margulis's claim that atmospheric oxygen was the driving force behind the evolution of eukaryotic cells. Perhaps mitochondria evolved later, after other important eukaryotic characteristics were already established. If substantiated, this new claim would be an important modification of Margulis's original theory.

Another interesting group of anaerobic protozoans includes those that have reinvaded such anaerobic environments as the sediments at the bottoms of ponds, lakes, and oceans. These protozoans have lost true mitochondria, but they contain organelles that are structurally similar to these respiratory organelles. These mito-chondrionlike structures now function in anaerobic metabolism. Because of their structural similarities, it seems likely that they evolved from mitochondria.

Still another anaerobic protozoan, the ciliate *Strombidium purpureum*, was discovered harboring bacteria capable of both photosynthesis and respiration. In the light, the ciliate avoids even trace amounts of oxygen, but it moves to areas where the light is optimal for bacterial photosynthesis. In the dark, the ciliate migrates to environments where small amounts of oxygen exist. The symbiotic bacteria use the oxygen to produce ATP by respiration. Here is a case of a single symbiotic bacterium playing the roles of both chloroplast and mitochondrion. This may be a biological oddity that does not closely resemble any of the early stages of cellular evolution. On the other hand, studying these present-day examples of endosymbiosis may provide useful models for understanding how eukaryotic organelles evolved over two billion years ago.

QUESTIONS AND ACTIVITIES

1. What does this case show about the following aspects of doing biology?
   — criticism and the revision of theories
   — piecemeal acceptance of new theories
   — resolution of scientific controversies
   — interrelationships of different scientific disciplines

2. Analogies are frequently used by scientists to justify new theories. For example, Margulis claims that hypothetical organisms from the past are somewhat similar to present-day cases of endosymbiosis (e.g., *Paramecium bursaria* and *Myxotricha paradoxa*). How convincing is each of Margulis's analogies? What other forms of evidence might be used to strengthen each analogy?

3. Why does Margulis believe that endosymbiosis was the result of adaptation to an aerobic environment? If this assumption turns out to be incorrect, how might it affect the scientific acceptance of the SET?

4. How might supporters of the SET and autogeny explain each of the following observations? Decide how strongly each observation supports the SET, autogeny, or both.

   a. Mitochondria and chloroplasts contain DNA molecules similar to those found in bacteria.

   b. The earliest prokaryotic fossils date from about 3.5 billion years ago, and the earliest eukaryotic fossils date from about 2 billion years ago.

c. Both chloroplasts and mitochondria contain two sets of membranes, an outer membrane and a greatly folded inner membrane.

d. Oxygen was not found in the primitive atmosphere, but was formed as a by-product of photosynthesis beginning about 2.5 billion years ago.

e. Both chloroplasts and mitochondria contain ribosomes smaller than typical eukaryotic ribosomes but about the same size as bacterial ribosomes.

f. Chloroplasts and mitochondria are about the same size as some bacteria.

g. Some cyanobacteria are found living inside eukaryotic organisms.

h. Mitochondria and chloroplasts are incapable of independent existence.

i. DNA and ribosomes have not been found in eukaryotic flagella.

j. Some cyanobacteria contain complex systems of internal photosynthetic membranes somewhat similar to those of chloroplasts.

k. Prior to cell division, chloroplasts and mitochondria reproduce by binary fission, just as bacteria do.

l. If mitochondria or chloroplasts are removed, the cell cannot regenerate them.

m. In some cases, spirochetes are attached to the plasma membranes of eukaryotic host cells. These spirochetes look and sometimes act like flagella.

n. Antibiotics that inhibit protein synthesis on ribosomes in bacteria also have this effect on mitochondrial ribosomes in eukaryotic cells. But the cytoplasmic ribosomes of eukaryotic cells are not inhibited by the antibiotics.

SUGGESTED READING

Darden, L., and N. Maull. 1977. "Interfield Theories." *Philosophy of Science* 44: 43–64.

Douglas, A. E. 1994. *Symbiotic Interactions*. Oxford, England: Oxford University Press.

Fenchel, T., and B. J. Finley. 1994. "The Evolution of Life without Oxygen." *American Scientist* 82: 22–29.

Maienschein, J. 1991. "Cytology in 1924: Expansion and Collaboration." In K. R. Benson, J. Maienschein, and R. Rainger, eds., *The Expansion of American Biology*. New Brunswick, NJ: Rutgers University Press.

Mann, C. 1991. "Lynn Margulis: Science's Unruly Earth Mother." *Science* 252: 378–381.

Margulis, L. 1993. *Symbiosis in Cell Evolution: Microbial Communities in the Archean and Proterozoic Eons*. 2nd ed. San Francisco: Freeman.

McDermott, J. 1989. "A Biologist Whose Heresy Redraws Earth's Tree of Life." *Smithsonian* 20(5): 72–81.

Sapp, J. 1994. *Evolution by Association: A History of Symbiosis*. New York: Oxford University Press.

Taylor, F. J. R. 1979. "Symbioticism Revisited: A Discussion of the Evolutionary Impact of Intracellular Symbioses." *Proceedings of the Royal Society of London* Series B 204: 267–286.

CHAPTER 4

Nettie Stevens & the Problem of Sex Determination

JOEL B. HAGEN

☐ INTRODUCTION

We may think that we know about the birds and the bees, but animal sexuality takes many unusual forms. Anemone fish, which live on coral reefs, begin life as males but later develop into females. In other coral reef fish, called wrasses, the sequence is reversed. A single male lives with several smaller females. If the male dies, the largest female takes his place. Within days she begins to produce sperm instead of eggs. In sea bass, each individual is both male and female. Although the sea bass has a combined testis and ovary, mating still occurs and an individual almost never fertilizes its own eggs. On the other hand, whiptail lizards, found in the southwestern United States, are always female. These lizards reproduce asexually through the development of unfertilized eggs. Sexual behavior would seem to be useless in this case, but whiptail lizards court and mate much like their sexually reproducing relatives. Before ovulation females behave like females, but after laying eggs they behave like males. Apparently male behavior still serves an important function in whiptail lizards, because isolated females lay fewer eggs than those who engage in pseudosexual intercourse with other females.

Unusual forms of animal sexuality challenge our commonsense notions of what it means to be male or female. These phenomena remind us that sexuality is a complex combination of anatomical, physiological, and behavioral characteristics. They also raise one of the most fundamental questions in biology: What determines sex?

Biologists have always been intrigued by this question, but sex determination generated particular interest around 1900. Perhaps this interest was partly due to the social climate of the times. The close of the Victorian era, "the sexless age," brought a more liberal attitude toward sexual matters. Important scientific advances also made sex determination interesting. The optical quality of microscopes had improved dramatically during the nineteenth century, allowing biologists to observe the nucleus, chromosomes, and other cellular structures with great clarity. By 1900 many biologists were convinced that studying cells—sperm, eggs, and zygotes—was the key to understanding sex. The rediscovery of Gregor Mendel's work on heredity added another new dimension to the study of sex. Was sex inherited according to Mendel's laws? How were patterns of inheritance related to the processes of cell division and fertilization? How did male and female characteristics develop in the embryo?

ALTERNATIVE THEORIES OF SEX DETERMINATION

At the turn of the century, several theories attempted to explain how individuals become male or female. None of these explanations was satisfactory. Supporting data were ambiguous, and biologists disagreed on what specific problems were most important. For example, some biologists were interested in how sex is inherited, while others were more interested in how sexual characteristics develop in the embryo. Not surprisingly, there were also major disagreements concerning research methods, basic assumptions, and philosophical implications of various theories. This situation might appear chaotic, but the very uncertainty surrounding sex determination appealed to ambitious biologists. With perseverance, hard work, and a bit of luck, a scientist might make a discovery of fundamental importance.

Before 1900, most biologists were "externalists." They believed that sex was caused by interactions between a sexually undifferentiated embryo and its external environment. As the embryo developed, it became male or female as a result of these interactions. How did this occur? Experimenting with tiny aquatic animals called rotifers, one influential biologist concluded that temperature was the controlling factor. At high temperatures most rotifers became males, while at lower temperatures nearly all became female. Other biologists claimed that nutrition influenced sex determination. Well-fed caterpillars almost always became females, but malnourished caterpillars usually became males.

Externalist theories were enormously popular around 1900. Not only were they supported by experimental evidence, but they also reflected the popular philosophical position that adult structures develop from scratch in the embryo (**epigenesis**). **Preformationism**, the rival belief that adult structures are already present in miniature form in the egg or sperm, had been widely rejected. It was perhaps natural, therefore, that many biologists believed that sex was undetermined in the zygote and gradually emerged as the developing embryo interacted with its surrounding environment.

Unfortunately for externalists, experimental results are usually open to more than one interpretation. Most of the externalists' experiments were conducted on whole populations rather than individual organisms. Externalists were, therefore, incapable of predicting whether any particular individual would become male or female. Critics quickly pointed out that altered sex ratios in populations could also be explained by selective mortality. The fact that, at lower temperatures, rotifer populations contained mostly females might simply mean that many embryonic males died in the cold. Critics also complained that almost any environmental factor seemed to affect sex ratios. Was there one environmental cause of sex determination, or many?

PROBLEM
Design a simple experiment deciding whether selective mortality is responsible for altered sex ratios in the populations of rotifers grown at two different temperatures.

The rediscovery of Gregor Mendel's work in 1900 stimulated considerable interest in studying patterns of inheritance. Some biologists, "Mendelian internalists," enthusiastically argued that sex was inherited according to Mendel's principles. Some Mendelians claimed that every gamete carried a sex-determining factor (what we would now call a **gene**). Each egg or sperm carried either a male or a female factor. The combination of factors in the zygote determined the sex of the offspring. Critics immediately pointed out that if this were true, then sex ought to be inherited according to Mendel's 3:1 phenotypic ratio. Mendelians responded by hypothesizing selective fertilization; certain combinations of gametes were more likely than others.

> ### *PROBLEM*
> **Suppose that sex is a simple Mendelian characteristic determined by two hereditary factors, one for maleness and one for femaleness. Based on this assumption, is it possible to account for the 1:1 sex ratio found in most species of animals? Can the hypothesis of selective fertilization help explain this sex ratio?**

Mendelian theory appealed to many biologists, but it had great difficulty explaining how sex could be inherited. An obvious weakness was accounting for the 1:1 sex ratio found in most species. If selective fertilization occurred, it ought to be observable, but sperm and eggs seemed to combine randomly. More serious, perhaps, the Mendelians could not adequately explain what the sex-determining factors were, where they were found in the cell, or exactly how they caused an individual to become male or female. Critics insisted that Mendelian factors must have a physical existence, but in 1900 this could not be demonstrated. The very idea of sex-determining "factors" smacked of preformationism, because it suggested that the sex of an egg, sperm, or zygote was set even before development began. This struck critics as a retreat to discredited theories of the past.

Some biologists ("non-Mendelian internalists") tried to find a middle ground between the externalists and the Mendelians. Like the externalists, they believed that sex determination must be understood as a gradual, developmental process. Also like the externalists, they were highly suspicious of the invisible Mendelian "factors." Unlike the externalists, however, these biologists looked for the causes of sex determination inside the embryo. Primarily trained as embryologists, they believed that complex physiological changes in the nucleus or cytoplasm caused embryos to become male or female.

Non-Mendelian internalism was particularly popular in the United States. It attracted prominent biologists such as Thomas Hunt Morgan and Edward Beecher Wilson. Both of these men later became champions of the theory of sex chromosomes, but in 1900 they were leading critics of Mendelism. They changed their minds about sex chromosomes partly as a result of important discoveries made by their student and colleague: Nettie Maria Stevens.

THE MAKING OF A CELL BIOLOGIST

The dawning of the twentieth century brought many new opportunities to aspiring professional women in the United States. Although they would not be guaranteed the right to vote for another two decades, women in 1900 were beginning to enter professions that had previously been closed to them. Talented women, routinely denied educational opportunities during the nineteenth century, could now pursue graduate training at some of the best universities in the nation. This opened new possibilities for careers in science, where jobs were becoming increasingly specialized. Women could pursue both research and teaching, although they continued to face many barriers to promotion and professional advancement.

The opportunities and constraints facing women in science during the early twentieth century are illustrated by the career of Nettie Stevens (Figure 4.1). As a young woman, Stevens taught high school and was a librarian, traditional occupations for unmarried women during the late nineteenth century. Her teaching duties included courses in physiology and zoology, as well as mathematics, Latin, and English. Her interest in zoology may have been encouraged by summer field biology courses that she took at the seashore near Martha's Vineyard during the early 1890s. In 1896, at the age of 35, Stevens entered college at Stanford, one of several new universities that admitted women. After receiving her B.A. in 1899 and her M.A. in 1900, Stevens left California to become a doctoral student at Bryn Mawr College.

It might seem odd that an aspiring scientist should choose a small women's college for graduate training, but Bryn Mawr was an excellent choice for Stevens. The biology department had developed a national reputation under the leadership of Edmund Beecher Wilson, perhaps the leading cell biologist in the United States. Although Wilson had moved on to Columbia University, he maintained close ties with his former department. Wilson's place at Bryn Mawr was taken by his friend, Thomas Hunt Morgan, who was already a prominent biologist, well-known for his studies on animal heredity and development (see Chapter 5).

Working with Morgan, who was five years younger than she was, Stevens enjoyed opportunities that women before her did not have. Although she collaborated with Morgan on some work, Stevens also conducted independent research. Through Morgan's influence, Stevens was able to spend two summers in Europe working in the

FIGURE 4.1 Nettie Maria Stevens. *Source:* The Carnegie Institution of Washington.

laboratory of Theodor Boveri at the prestigious Naples Zoological Station. At the time, Boveri was one of the world's leading experts on chromosomes. Stevens also went to work at the Marine Biological Laboratory at Woods Hole on Cape Cod, where Morgan, Wilson, and other leading American biologists spent their summers doing research.

By the time she completed her Ph.D. in 1903, Stevens was a seasoned scientist. She had already written several scientific papers for leading biological journals. The instructorship at her alma mater, which she took immediately after graduation, involved heavy teaching responsibilities, but it provided a laboratory for her research. Fortunately, Stevens won research grants from one of the most prestigious scientific foundations in the United States: The Carnegie Institution of Washington. This financial support allowed her some free time to pursue research directed at one of the great biological questions of the day: how are chromosomes involved in sex determination?

CHROMOSOMES, ACCESSORY CHROMOSOMES, AND SEX CHROMOSOMES

Today we take for granted that chromosomes carry the units of heredity (genes), but establishing this fact was not a straightforward process of discovery. The term **chromosome** was first used in 1888 to describe the tiny, threadlike structures in the nucleus that many cell biologists had studied during the preceding decade. The movements of chromosomes during cell division and their significance for both sexual and asexual reproduction became major scientific problems. By 1900, the processes of mitosis, meiosis, and fertilization had been accurately described in both plants and animals. There was considerable disagreement about the function of chromosomes, however. If you reflect upon our current understanding of how chromosomes work, you will realize that very little of this knowledge comes from direct observation. Thus the development of a satisfactory theory of chromosomes involved considerable speculation, as well as the piecing together of fragmentary evidence. Two controversial hypotheses were particularly important for guiding Nettie Stevens's research.

Biologists knew that chromosomes usually come in pairs, the members of which have the same size, shape, and placement of centromeres. During the late 1890s, cell biologists discovered that unpaired chromosomes are also sometimes found in cells. What we now call the X chromosome was referred to as the "accessory chromosome," and in 1902 one of Wilson's students, Clarence E. McClung, implicated it in sex determination. Thinking that it was found only in some sperm (and never in eggs), McClung speculated that if a zygote received the accessory chromosome, it became male; if not, it became female (Figure 4.2). This, of course, turned out to be incorrect, but it was an important hypothesis for two reasons. First, it stimulated considerable interest in studying the relationship between chromosomes and sex. Second, in suggesting that there are two types of sperm, McClung was on the right track.

A broader speculation was made by another of Wilson's students, Walter Sutton, and independently by the German biologist Theodor Boveri. According to the Sutton-Boveri hypothesis, chromosomes maintain their individuality and physical integrity even when they are not visible. This was a controversial claim because many biologists believed that chromosomes formed more or less randomly before each cell division. Sutton also emphasized the striking parallel between Mendel's law of segregation, which applied to hereditary factors, and the separation of chromo-

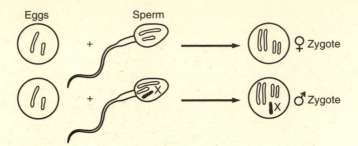

FIGURE 4.2 Clarence E. McClung's hypothesis of sex determination by accessory chromosomes. McClung claimed zygotes with an accessory (X) chromosome became males; those without the accessory (X) chromosome became females. Later studies refuted McClung's hypothesis.

some pairs during meiosis. Did chromosomes carry Mendelian factors? If so, did the accessory chromosome carry a sex-determining factor?

Both Mendelians and non-Mendelians studied chromosomes, and in the years following 1900 both groups realized that these nuclear structures were important for understanding sex determination. The two groups, however, tended to interpret the results of their studies differently. Mendelians claimed that chromosomes *caused* heredity, because they carried the hereditary factors postulated by Mendel. Because it carried the Mendelian factor for sex, the "accessory chromosome" was really a "sex chromosome." Non-Mendelian internalists such as Morgan and Wilson were more cautious, admitting only that chromosomes were somehow *correlated* with heredity. Perhaps it was the case that the accessory chromosome simply acted as a "marker." The chromosome indicated the sex of an individual, but it did not actually cause the individual to be male or female. In 1903, when Nettie Stevens earned her Ph.D., the exact nature of chromosomes and their role in sex determination remained open questions.

STEVENS'S STUDIES ON SEX DETERMINATION

The uncertainty over sex determination is borne out by Stevens's early research. Together with Morgan, she did experiments on aphids to test the claim that temperature alters sex ratios in populations. These experiments failed to confirm the externalists' hypothesis. In her early cellular studies of aphids (which she did by herself), Stevens also failed to detect McClung's accessory chromosomes. Thus, at the end of 1904, she concluded that although it seemed likely that sex was somehow determined by the eggs and sperm, exactly how it was determined remained unclear.

During the next two years, Stevens completed a comparative, cellular study of several species of insects, drawn from a diverse group of beetles (Coleoptera), butterflies and moths (Lepidoptera), and true bugs (Hemiptera). This was painstaking work. First she dissected the tiny gonads from the insects. These were fixed in a preservative solution, embedded in paraffin blocks, and sliced into very thin sections. The tissue sections were then mounted on microscope slides and stained with one of several dyes. Careful observations of the tissue sections revealed gametes in various stages of development. If a cell had been just beginning to divide and the tissue had been cut at just the right angle, all of the chromosomes were clearly visible and could be accurately counted.

In the common mealworm, *Tenebrio molitor*, Stevens found that the body cells of females contained 20 large chromosomes, while body cells in males contained 19 large chromosomes and 1 small chromosome (the Y chromosome). She also found that although mealworm eggs always contained 10 large chromosomes, there were two types of sperm: 50 percent contained 10 large chromosomes, 50 percent contained 9 large chromosomes plus the 1 small Y chromosome. This pattern was found in the other species that she studied except that, in a few species, the small (Y) chromosome was completely absent. In these cases, 50 percent of the sperm contained a set of chromosomes identical to that found in eggs, while in the other 50 percent the number of chromosomes was one fewer than that found in eggs. Stevens's results could be generalized as follows: for any given species, all eggs are the same, but there are two distinct types of sperm.

In her conclusion, Stevens pointed out that a male is produced whenever an egg is fertilized by a sperm carrying the small Y chromosome (or lacking this chromosome); a female is produced whenever an egg is fertilized by a sperm carrying the large X chromosome (Figures 4.3(A) and 4.3(B)). Having stated this conclusion, however, she refused to claim that the X and Y chromosomes could accurately be described as "sex chromosomes." These chromosomes seemed to play some hereditary role in sex determination, but how they did so was still unclear. In other words, Stevens had found an important correlation between chromosomes and sex determination, but she couldn't prove causation. She cautiously concluded that further evidence was needed before one could speak confidently of "sex chromosomes."

With the benefit of hindsight, Stevens's conclusion seems curiously conservative, but other prominent biologists were even more hesitant about drawing general conclusions from a relatively small sample of data. Wilson made similar observations of chromosomes and published his results at about the same time Stevens did. Like Stevens, he refused to endorse the idea of sex chromosomes. Although he admitted that chromosomes provided the best "working hypothesis" for explaining sex determination, he remained open to the possibility that they were simply indicators, rather than determiners of sex. Morgan was even more reluctant to accept the idea of sex chromosomes. Some of Stevens's contemporaries, particularly embryologists, never acknowledged that chromosomes might play a role in sex determination.

THE DISCOVERY OF SEX CHROMOSOMES

Looking back at Stevens's contribution to science after her death in 1912, Thomas Hunt Morgan claimed that biologists had been too cautious about sex chromosomes. But Morgan, and to a lesser extent Stevens and Wilson, were among this conservative group. Part of their conservatism stemmed from the need to confirm Stevens's hypothesis with data from other species. This turned out to be a vexing puzzle. In some groups (birds and butterflies), it turned out that eggs were heterogametic and sperm were homogametic (Figure 4.3(C)). The development of unfertilized eggs was

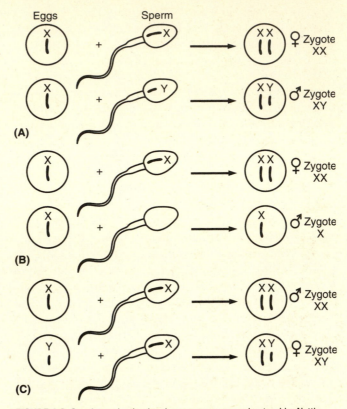

FIGURE 4.3 Sex determination by chromosomes as understood by Nettie Stevens and her contemporaries. (A) The XX, XY system found in mammals and some of the insects studied by Stevens. (B) In some other insects there is no Y chromosome; females have two sex chromosomes (XX), but males have only one (X). (C) Other biologists later discovered that in birds and some insects, the male is XX and the female is XY. Some cell biologists used the symbols WW and WZ to distinguish this from the XX, XY system.

also a major problem to be explained. In some species that reproduced asexually, unfertilized eggs developed into females, but in other species they developed into males. Numerous examples of individuals that had both male and female characteristics, also had to be explained. All of these pieces of the puzzle needed to fit together before many biologists accepted the idea of sex chromosomes.

From a historical perspective we can see that the "discovery" of sex chromosomes was not so much a single event as a period of transition in ideas about sex determination. In 1902 Clarence McClung first claimed that sex chromosomes existed, but the hypothesis that he put forward to explain sex determination turned out to be incorrect. During the next ten years, several biologists studied the accessory chromosomes in many species of animals and tested alternative hypotheses about sex determination. As a result, they gradually changed their views on how sex is determined. This was a major shift, for it meant changing the type of problem studied, the methods used, and some basic philosophical assumptions. Although biologists did not completely abandon the goal of explaining how sexual characteristics develop in the embryo, interest shifted to the problem of how sex is inherited. Stevens, Morgan, Wilson, and other biologists also had to set aside some of their fundamental philosophical commitments—for example, their fear that Mendelism would lead scientists backward to discredited preformationist ideas. As they made this intellectual shift, they also had to embrace a new set of scientific methods, particularly the experimental breeding of fruit flies (*Drosophila*) and other organisms. In 1914 Morgan was finally ready to present a comprehensive theory of sex determination by chromosomes in his popular book, *Heredity and Sex.*

Tragically, Nettie Stevens died of breast cancer in 1912, just when the idea of sex chromosomes was becoming well established. In the years that followed, her contribution to the discovery was often viewed as that of a data collector, whose careful observations were used by others (Morgan and Wilson) to create the chromosome theory of sex determination. This interpretation has now been rejected by most historians, who have pointed out her important theoretical contributions to this discovery. Several years before Morgan and probably several weeks before Wilson, she cautiously proposed an explanation for sex determination by chromosomes. For the rest of her short career, she continued to gather evidence to support this theory. Together with her two male colleagues, Stevens played a critical role in the discovery of sex chromosomes.

☐ *EPILOGUE*

This case study might be interpreted as the victory of Mendelism over two rival theories. It is well to remember, however, that throughout his life Morgan held out hope that a single theory would explain both the inheritance and development of sex. With the rise of molecular biology, studies of heredity and development have finally converged much as Morgan had hoped that they would. For example, scientists have recently discovered a male sex-determining region (SRY) on the Y chromosomes. The specific genes making up this SRY region, the proteins for which they code, and their developmental functions are important problems in current research.

We are also increasingly aware that being female or male involves more than simply the possession of XX or XY chromosomes (or the genes that they carry). Pieces of chromosome are sometimes lost or translocated to other chromosomes. As a result, we now realize that there are some XX males and XY females. Sexuality, it now appears, is a complex phenomenon involving sex organs, secondary sex characteristics, and behaviors that can only partly be explained in terms of genes and chromosomes. The physical, biological, and social environment also plays a crucial role.

If the environment is important, what about the discredited externalist explanations of sex determination? Early theories of environmental sex determination were based upon poorly controlled experiments and were rightly rejected by biologists. Perhaps there was more than a grain of truth in these incorrect ideas, however. Textbooks often omit this fact, but not all animals have sex chromosomes. Other genetic systems exist, and in some of these cases the environment plays an important sex-determining role. For example, in many species of reptiles environmental factors such as temperature act as developmental "switches." When eggs are incubated at some temperatures, males are produced; at other temperatures, females are produced. Temperature seems to influence the production of important enzymes and hormones in the developing embryo. Discovering exactly how this happens will require a better understanding of heredity, development, and the environment.

QUESTIONS AND ACTIVITIES

1. What does this case show about the following aspects of doing biology?
   — uses and limitations of indirect evidence
   — resolution of scientific controversies
   — revision of scientific theories
   — positive role of incorrect hypotheses
   — gradual versus sudden discoveries

2. Why is it difficult to identify a single scientist as the discoverer of sex chromosomes? How did each of the following scientists contribute to this discovery: McClung, Sutton, Stevens, Wilson, Morgan?

3. How was McClung's hypothesis about accessory chromosomes incorrect? Can you think of a plausible explanation for his mistake? Why are incorrect hypotheses, like McClung's, sometimes very important in science?

4. The Sutton-Boveri hypothesis cannot be confirmed solely on the basis of microscopic observation. Were biologists in 1903 justified in accepting the hypothesis even without conclusive proof? Why was the hypothesis necessary for the theory of sex chromosomes?

5. In her cellular studies, Nettie Stevens concluded that the X and Y chromosomes were correlated with sex but did not necessarily cause sexual differentiation. What is the difference between correlation and causation? What other forms of evidence might have been used to demonstrate a causal relationship between chromosomes and sex?

6. Consider the sea bass, which is both male and female, or reptiles whose sex is partly determined by temperature. If biologists had known about these unusual examples in 1905, how might this information have influenced the acceptance or rejection of the sex chromosome theory?

SUGGESTED READING

Allen, G. E. 1966. "Thomas Hunt Morgan and the Problem of Sex Determination." *Proceedings of the American Philosophical Society* 110: 48–57.

Badge, R. L. 1991. "SRY and Sex Determination." *The Journal of NIH Research* 3: 57–59.

Brush, S. G. 1978. "Nettie M. Stevens and the Discovery of Sex Determination by Chromosomes." *Isis* 69: 163–171.

Crews, D. 1994. "Animal Sexuality." *Scientific American* 270(1): 108–114.

Darden, L. 1991. *Theory Change in Science: Strategies from Mendelian Genetics.* Oxford, England: Oxford University Press.

Farley, J. 1982. *Gametes and Spores: Ideas about Sexual Reproduction 1750–1914.* Baltimore: Johns Hopkins University Press.

Maienschein, J. 1984. "What Determines Sex? A Study of Converging Approaches, 1880–1916." *Isis* 75: 457–480.

Ogilvie, M. B. 1991. "The 'New Look' Women and the Expansion of American Zoology: Nettie Maria Stevens (1861–1912) and Alice Middleton Boring (1883–1940)." In K. R. Benson, J. Maienschein, and R. Rainger, eds., *The Expansion of American Biology.* New Brunswick, NJ: Rutgers University Press.

Ogilvie, M. B., and C. J. Choquette. 1981. "Nettie Maria Stevens (1861–1912): Her Life and Contributions to Cytogenetics." *Proceedings of the American Philosophical Society* 125: 292–311.

CHAPTER 5

Thomas Hunt Morgan & the White-eyed Mutant

DOUGLAS ALLCHIN

☐ INTRODUCTION

In 1910, the lively ragtime music of Scott Joplin was sweeping the nation. Harry Houdini amazed crowds with his spectacular escapes from straitjackets and locked trunks. Cities were booming. Major industry had increased dramatically in the late 1800s, and with it the demand for more workers to operate factories. Many people had moved from rural to urban areas to find jobs. In New York City, residents were accustoming themselves to this growth and to a new underground train system, a subway, opened six years earlier to help accommodate their needs. City residents were also still fascinated with "aeroplane" flights. The Wright Brothers had been the first to fly at Kitty Hawk, North Carolina, only seven years earlier. *The World*, one of the city's newspapers, capitalizing on the excitement surrounding manned flights, had offered a $10,000 prize for the first person to fly from Albany to New York non-stop.

On the upper west side of New York City, on the sixth floor of Schermerhorn Hall on the campus of Columbia University, was a 16- by 23-foot room that would soon become renowned worldwide (Figure 5.1). It was Thomas Hunt Morgan's lab—or what passed for his lab. It was crowded with eight large desks, including those for undergraduates involved in his research. Papers were piled on the desks, shelves were filled with bottles, and the air smelled of yeast and fermenting bananas. Morgan (Figure 5.2) had gained widespread recognition for his experiments in embryology. He was also noted for his vigorous skepticism, especially about the ideas of Darwin and Mendel.

Morgan was pursuing many research projects. In one study on experimental evolution using fruit flies, he had recently encountered a white-eyed male. In a population of flies whose eyes were normally red, the white-eyed individual was certainly remarkable. Granted, individuals with unusual traits—"sports of nature"—did appear occasionally. But the traits often were lost again in subsequent generations. If the white-eyed trait was inherited, however, it might reflect a mutation—a significant, long-term genetic change in the population. If so, it held for Morgan a potential clue about how new species could originate. At the same time, the trait exhibited a strange and equally remarkable pattern of inheritance. So far, only males had been found with the white-eyed trait. Why? What feature of development, heredity, or species formation could account for this?

FIGURE 5.1 Thomas Hunt Morgan's lab in Schermerhorn Hall, Columbia University, New York (around 1920). Calvin Bridges, who originally started working with Morgan by washing bottles as an undergraduate, sits at his desk. *Source:* Courtesy of the American Philosophical Society Library, Stern Papers.

FIGURE 5.2 Thomas Hunt Morgan at work in his office (sometime between 1915 and 1920). *Source:* American Philosophical Library, Columbia University Department of Biology Photograph Collection.

THOMAS HUNT MORGAN AND THE LIFE SCIENCES, 1900–1910

Morgan's background influenced how he perceived the problem of the white-eyed fly. Most of Morgan's research had been related to a central question in biology at the time: What determines the form of an organism? The question had three parts: (1) How does a complex organism *develop* from a single cell and acquire its specific form? (2) How does an organism *inherit* from previous generations specific traits or features of its form? (3) How do new forms or species *evolve* on a larger historical time scale? The challenge throughout the 1800s had been to explain all three processes at once. Each question posed ways for Morgan to think about his white-eyed mutant: How did the new eye color originate? Was it an accident of embryological development? Could it be inherited? Did it represent a trait distinguishing a new species?

Darwin, of course, had provided one solution to the puzzle of organismal form: gradual evolution through natural selection. But his views were not uniformly accepted, even by biologists working after 1900. Morgan strongly criticized Darwinism in 1903. Selection, he noted, was only a negative process. It edited or reduced variation. How did new traits emerge in the first place? Biologists needed to search for the causes of variation that would explain how a new lineage evolved. In addition, Morgan did not see the concept of natural selection as rigorous or testable. Morgan's criterion of "scientific" proof was experiment in the laboratory. No one had demonstrated evolution or the creation of a new species in a lab (though Morgan might have been impressed with later studies by H. B. D. Kettlewell—see Chapter 1).

The premier area of study in biology at the turn of the century was development—how a fertilized egg is dramatically transformed into an embryo and how an embryo is further transformed into a complex adult organism (see also Chapter 4). When Morgan encountered the white-eyed mutant, he had already studied development extensively. For example, he had examined how the developing organism's form is affected by the initial orientation of the egg and by the concentration of salts in the fluid surrounding the egg. He had looked at how egg fragments (rather than whole eggs) developed when fertilized. Morgan also had a strong interest in how sex was determined. What made one organism male, another female? He had summarized his views in a 1909 research paper. Morgan's previous research offered several ideas for viewing the white-eyed trait in terms of development.

Despite the long-standing focus on development, the study of inheritance was expanding rapidly in the early 1900s. The research was closely linked to agriculture. As cities grew and population increased, demand for food rose. At the same time, because people had migrated to the cities, there were fewer hands to manage the farms. Farmers needed to increase crop yields. They turned to fertilizers, new methods of plowing, and better animal nutrition. But as U.S. Secretary of Agriculture James Wilson noted in 1910, these improvements based on the "environment" relied on costly expenditures every year. Farm profits could be vastly increased, he argued, by changing the plants themselves. Genetic improvement would be more permanent and, ultimately, less costly. Both public and private groups recognized the opportunity and invested heavily in agricultural research. Money flowed from the U.S. Department of Agriculture, from state agricultural stations, and from private agencies such as the Rockefeller Foundation, the Kellogg

Foundation, and The Carnegie Institution of Washington. With more funds available, research on heredity accelerated.

The rediscovery in 1900 of Gregor Mendel's work on inheritance in pea plants sparked new research. Mendel's notions about hybrids suggested how to search for better crops and domesticated animals. Some researchers thought that they might be able to modify the Mendelian factors of inheritance and began to look for them at the level of the cell. Morgan, however, was as skeptical of Mendelism as he was of Darwinism. Mendelians believed that heredity was controlled by unit particles, with one trait dominant over another. Morgan saw this as flying in the face of the subtle variety that he observed in organisms. Given his own studies on how the environment could influence development, Mendelism seemed too rigid. More importantly, Mendel's "factors" were unobservable. Many of the complex patterns of inheritance could be explained instead by the selective fertilization of gametes; that is, some gametes were more likely to be fertilized than others. Morgan considered Mendel's concepts too simple for the complexity of organisms that he knew well.

FRUITFUL FLIES

Morgan reportedly remarked (more than once) that he had done three kinds of experiments: "those that were foolish; those that were damn foolish; and those that were worse than that." He might well have used any of these phrases to assess one line of his investigations in the four years leading up to 1910. Beginning in 1906, Morgan began trying to induce the mutations that he thought fueled evolution. He tried various treatments: salts, sugars, acids, and alkali—the kinds of factors that he knew from his previous research could influence embryological development. If they could change the form of one organism, he reasoned, then they might be able to change the form of the entire species' lineage. But Morgan was not having much success. His failure was hardly due to foolishness, though—just bad luck.

While his search for mutants continued, Morgan was busy with other projects as well. He was thinking about the problem of sex determination. He was looking at regeneration in embryos. At the same time, like any university professor, he was also teaching classes.

Morgan's classes typically involved an independent research project, as was customary at the time. In 1907, a graduate student in one of his courses, Fernandes Payne, planned to study whether he could gradually induce blindness in organisms living several generations in the dark. Morgan would have endorsed such a study, given its experimental approach to an evolutionary question. The critical problem for a one-year course was time: how could he examine enough generations to make the study worthwhile? Payne's teacher at college had worked with fruit flies (also called vinegar flies, pomace flies, and sometimes banana flies), and they seemed an appropriate organism to study. He would be able to study at least ten generations in 9 months. Payne found nothing remarkable in his experiments, but Morgan had found a new research organism. The following year, Morgan began to incorporate fruit flies into his own work on mutations and experimental evolution.

One may wonder why Morgan would have found fruit flies worth researching. They provided no obvious clues to human physiology. Nor were they one of the agricultural organisms currently the target of so much research on genetic improvement. But fruit flies have great advantages for researchers. They do not take up much space—something to consider when your lab is small. They can be easily collected from the wild, even in the city. Caring for them is cheap and easy—an advantage when you have a busy schedule. In fact, Morgan kept the flies in half-pint milk bottles that he "borrowed" from the Columbia University cafeteria. The flies fed on yeast that grew on bits of overripe banana. The fly larvae also needed a surface on which to pupate, and so Morgan folded leftover envelopes and inserted them in the milk bottles. All in all, Morgan invested little time, money, or trouble for his "foolish" experiment. Fruit flies may have seemed trivial, but the choice made practical sense.

Morgan tried to induce mutations in fruit flies with the same treatments he had been using—salts, sugars, acids, and alkali. He exposed them to radium. A year later, he began subjecting large populations of flies to intense selection pressure, hoping to amplify the effect of mutations. Morgan was not having much success. In early 1910, when an old colleague visited his lab, Morgan waved his hand at the rows of bottles on the shelves, exclaiming, "There's two years of work wasted. I've been breeding these flies for all that time and have got nothing out of it."

A PUZZLING WHITE-EYED MUTANT

But soon thereafter, Morgan began to find flies with unusual traits. Some had a different coloration, others had a different body shape. In May, Morgan found a male with anomalous white eyes (Figure 5.3). Perhaps he was unveiling the clues to understanding evolution. The change in eye color was not dramatic enough to mark a new species. Still, it was outside the range of normal variation. Morgan had to ask whether each trait was a "sport of nature"—one of those random variants that was not inherited. Morgan wanted to know how the white-eyed trait, like others, would fare in future generations. So he bred the individual with sister females from the same population.

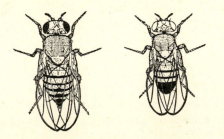

FIGURE 5.3 Mutant fly. Normal red-eyed female on left, white-eyed male on right. Note the different shapes and colors of the abdomens, which allowed Morgan and his students to determine the sex of each fly. *Source:* Thomas Hunt Morgan, *The Theory of the Gene* (Yale University Press, 1926), p. 60.

The white-eyed trait did not become more prevalent in the offspring. Nor did it generate eyes that were an intermediate pink, a mixture of half-red and half-white (color blending had occurred in some flower crosses). Instead, Morgan puzzled over the following results:

red-eyed female(s) x white-eyed male:

1,237   red-eyed (male and female)
   3   white-eyed (all male)

Three white-eyed males appeared among a large sample of over a thousand (a proportion of less than one-half of 1 percent). Morgan had plenty of options to consider. Should the three white-eyed offspring be viewed as having received the trait from the white-eyed parent? If the white eye was due to a lack of pigment, then could the small number reflect a dilution in the concentration of the pigment chemical in the sperm cell? Did the three males indicate that the trait would eventually proliferate throughout the population and form a new species or subspecies? Or were the three flies merely "sports of nature," perhaps like the original white-eyed male in the previous generation?

The trait appeared lost in the first generation. But the appearance of traits from generations earlier than one's immediate parents was a well-documented hereditary phenomenon. Morgan allowed the first-generation offspring to breed among themselves. He found the trait reappeared much as Mendel would have predicted(!):

F_1 red-eyed females x F_1 red-eyed males:

2,459   red-eyed females
1,011   red-eyed males
 782   white-eyed males

Morgan interpreted this as a Mendelian three-to-one ratio, though the ratio was not perfect. (Would you agree with his interpretation?)

Morgan's attention, however, was drawn by a distinctly non-Mendelian feature of the results, evident when the flies were sorted and counted by sex. As Morgan emphasized when he published his results, *"No white-eyed females appeared."* That is, *all the white-eyed flies were male.* What could possibly yield this Mendelian ratio split so dramatically and discretely across sexes?

Even though Morgan was looking primarily at evolution, his findings introduced questions about sex determination. Like many researchers encountering unexpected results, Morgan needed to draw on knowledge in a related field by talking with colleagues who might know something or by reading the published scientific literature. In this case, Morgan could "consult" himself, since, coincidentally in this case, he had researched sex determination.

The question of how sex was determined was still largely unsolved for Morgan. He was certainly aware of the work of his former graduate student, Nettie Stevens (Chapter 4), who claimed that sex was determined by the presence or absence of a Y chromosome. In fruit flies, she had observed that XX chromosome pairs occur exclusively in females, while XY pairs or X singles occur only in males. Stevens's views were shared by Edmund Wilson, chairman of the Zoology Department at Columbia, whose office was just down the hall from Morgan's. Morgan and Wilson were close friends and undoubtedly exchanged thoughts on matters such as this. But Morgan had interpreted Stevens's results slightly differently. In 1907, Morgan had suggested that sex was determined quantitatively through biochemical reactions of the chromatin—the material that made up the chromosomes. That is, above some threshold amount of chromatin, the zygote would become a female (or male), while below that threshold, the opposite sex would develop. Individuals with an XX pair of chromosomes would have more chromatin than X or XY individuals, and hence be one particular sex. Morgan had recently reiterated this view in a long 1909 publication.

Morgan would have been troubled by the implications of Stevens's hypothesis for his white-eyed fly. If sex was indeed determined by chromosomes, and the eye-color trait was related to sex, then traits like eye color must themselves be related to the chromosomes. Yet there were far more traits than there were chromosomes, and far more combinations of traits than combinations of chromosomes. Further, to imagine that material units carried predefined traits, as Mendelians claimed, implied that organisms were predetermined. Morgan knew, however, how the environment also affects growing organisms. If Morgan was going to pursue chromosomal notions of sex, he would have to rethink his own notions about sex determination and Mendelian inheritance.

Morgan could thus apply several ideas from his experience to the problem of the white-eyed mutant. The answer would not announce itself, however. He had to carefully think through the alternatives, which included:

—Sex and white eye result from some gametes being fertilized while others are not (selective fertilization).
—The white-eyed trait appears only in association with the Y chromosome in males.
—The eye-color trait is associated with the X and Y chromosomes in some other way.
—Sex and white eye are each determined by a quantitative biochemical threshold of chromatin or some other compound.

PROBLEM
Select the hypothesis you think is most plausible. Review how it could explain the available evidence. Suggest an investigation that would allow you to further confirm or demonstrate the explanation. Identify the results you would expect. Would your prospective investigation allow you to rule out other plausible explanations?

Consider using diagrams or physical objects to help organize your thinking.

A RELUCTANT CONCLUSION?

Morgan ultimately concluded that the white-eyed trait was coupled to a sex factor similar to the X chromosome. At first, this may not seem plausible, because only males had exhibited the white-eyed trait, while females also had X chromosomes but did not exhibit the trait. To reason through this, you have to track chromosomes and eye color simultaneously (though Morgan himself continued to think in terms of X and Y factors, not chromosomes). The original white-eyed mutant, Morgan concluded, had the white-eyed factor, *W*, *coupled* with the X factor for sex. The trick was to follow the coupled X–*W* *pair* of factors through the generations and know when the white-eyed trait would appear.

In detail, the original white-eyed male would be XY but also have an X–*W* couple. When he produced gametes, the X and Y sex factors would separate, and half the offspring would receive the coupled X–*W* trait. Which half? Because the other parent, an XX female, could contribute only an X factor (or chromosome), the offspring with the father's coupled X–*W* would all be XX, or females. But the trait would not be expressed because these females would also have inherited a dominant red-eyed trait along with the X factor from the red-eyed mother. In other words, red-eyed traits would also be coupled to the X factor, making X–*R* couples. The female offspring would be X–*W*, X–*R*.

When the first-generation offspring mated with each other, the X–*W* couple (now in the F_1 females) would combine again with a Y factor from the F_1 males. The result, X–*W* and Y, would be a white-eyed male. (Because there was no dominant red-eyed factor, the males would be white-eyed.)

Why were there no white-eyed females? Some F_2 females would indeed have the X–*W* couple. But they also would have inherited an X–*R* couple from an F_1 male. They would be female, XX, but red-eyed, *RW* (with *R* dominant). Overall, you could trace the traits just as you would in a Mendelian cross. The key was to treat eye color and the sex factor (or chromosome) as a single unit, not two.

PROBLEM
At this point Morgan had found only white-eyed males. Were white-eyed females possible according to his hypothesis? If so, predict a cross that would yield such a female.

For further guidance, consider: What would be the hereditary composition of a white-eyed female? Work backwards to determine a combination of gametes that might create a white-eyed female. Which parents from Morgan's cultures would produce such gametes?

CHALLENGE
What would a white-eyed female indicate about the explanations based on biochemical thresholds or selective fertilization?

Morgan's results indeed fit elegantly with the chromosome hypothesis of sex determination and with the chromosome hypothesis of Mendelian factors. But Morgan was not thoroughly convinced. As noted earlier, Morgan had exceptionally high standards of experimental evidence. In addition, you might imagine that he would not easily have abandoned the objections to Mendel that he had held for so many years.

Not long afterwards, however, a mutant with a yellow body appeared. Alfred Sturtevant, one of Morgan's students, showed that the yellow-body trait was inherited the same way as the white-eye trait. Moreover, the inheritance of the two traits could be "coupled" (later, "linked") with each other, not just with the sex factor. Here was evidence that two traits seemed to be carried together by the same (X) chromosome. The results would have answered Morgan's earlier objection that a chromosome could not carry more than one trait. The chromosomal theory of inheritance would also explain why the eye-color factor and sex factor did not segregate independently from each other during reproduction: they would be literally linked together chromosomally. You might imagine Morgan's somewhat reluctant conclusion: Mendel's scheme worked, after all, and sex and other traits were determined by the chromosomes. Geneticist William Bateson described Morgan as having "a thick head," but how could Morgan have dismissed the evidence from experiments in his own lab?

☐ *EPILOGUE*

In retrospect, the appearance of the white-eyed mutant was a significant turning point for Morgan—and for biology as well. Morgan soon shifted the focus of his research to the immediate problems presented by the white-eyed fly: genetics. And it remained there for the next two decades. Morgan not only accepted Mendelism but became one of its strongest proponents. The fruit fly offered a productive avenue of research for him and his students. They capitalized on the many funds available for agricultural research, and the subsequent work of the "Morgan Group" was funded almost exclusively by the Carnegie Institution. The office became even more crowded as jars of fruit flies accumulated on the desks and shelves. The room was often noisy with chatter. The otherwise modest office became a center of genetics research known internationally simply as "the Fly Room." The humble fruit fly also gained renown. Even now, the fruit fly is one of the most studied organisms in biology, along with humans, *E. coli* bacteria, white mice, Rhesus monkeys, and the roundworm *C. elegans*.

In their subsequent work, Morgan and his students extended the evidence that genetic information is located on the chromosomes. This has since been recognized as their most significant contribution. In addition, they established that each gene has a specific position on the chromosomes. They argued that genes were arranged linearly and that their relative positions or distances could be "mapped." Their proposals generated a spirited debate. By 1926, however, Morgan was able to summarize the resolution of these controversies in a now landmark book, *The Theory of the Gene*. He was awarded a Nobel Prize in 1933 for leading the work linking chromosomes and heredity.

Morgan's research had traveled a quite unexpected path. The work first focused on experimental evolution with fruit flies, led through questions about development and sex determination, and then passed into the nature of genetics at the cellular level, where it spawned a major and enduring field of research. During this transition, views held at the turn of the century had also changed dramatically. Morgan had answered many questions about inheritance without also answering questions of development and evolution.

In the decades that followed, several biologists who studied under Morgan made significant discoveries in both these areas. Theodosius Dobzhansky, for example, studied fruit fly populations in the wild, documenting the relationship among species chromosomally. His conclusions about evolutionary change are still central today. George Beadle, another of Morgan's former graduate students, teamed up with Edward Tatum to study how genes were expressed biochemically in the cell—

a key process for understanding development. After working with fruit flies for a while, they switched to studying a mold and later introduced the notion that each gene is associated with a specific enzyme or protein. In other words, genes become traits through a functional protein or enzyme.

Nearly a century after Morgan, biologists are beginning to integrate the fields of genetics, evolution, and development more completely—and the fruit fly is once again central. Recent studies have focused on a gene identified by Morgan's group, named *eyeless*. In mutant flies that lack the normal gene, the eye is absent or partial. Researchers have now identified the developmental significance of the gene. It appears to be a regulatory or control gene. That is, it triggers the expression of all the other genes that lead to the development of the eye. When it is absent, no eye develops. The *eyeless* gene also appears to be similar to a corresponding gene in mice, suggesting that fruit flies and mice have a common evolutionary origin. Researchers further explored this evolutionary relationship by transplanting the mouse's gene into fruit fly larvae, where it induced the development of a *fruit fly* eye (not a mouse eye) on a leg! Such spectacular cross-species gene transplants hold clues to understanding how regulatory genes act in development and how changes in them may lead to evolution. Biologists are at last developing a comprehensive explanation that unifies evolution, development, and heredity. And the fruit fly is once again proving to be a "fruitful" organism for study.

QUESTIONS AND ACTIVITIES

1. What does this case show about the following aspects of doing biology?
   — chance or accident
   — the role of theoretical perspective in interpreting evidence
   — the posing of problems
   — the cost of research and funding research

2. In his published paper, Morgan characterized his three white-eyed males in the first-generation offspring as "due evidently to further sporting." Was this an adequate explanation? How fully does a researcher have to address all his or her results in a formal scientific report? Did Morgan have a responsibility to report those results at all, since they were incidental to his conclusions?

3. The case of the white-eyed mutant highlights how chance events can dramatically shift the direction of research. How should this affect how research is funded? If you were a science policy maker, how would you decide which specific researchers or research projects to fund?

4. According to Morgan's initial hypothesis, the F_2 females would have differed in hereditary makeup, even though they all had red eyes. How would you demonstrate whether the individuals that appeared the same possibly had different genetic compositions?

5. Morgan's original system of notation (which he later revised) allows us to glimpse his thoughts as they were still being formed. In his original publication, he described the white-eyed mutant as WX–W and the red-eyed wild female as RX–RX. Here, he followed the XX/X pattern of sex rather than the XX/XY pattern. He also assumed that all individuals had two Mendelian factors for eye color, so that males had an uncoupled factor for eye color. He described the first two crosses as follows:

WX–W (male)
RX–RX (female)

RWXX (50%)   RWX (50%)
Red female   Red male

and

RX–WX (F_1 female)
RX–W (F_1 male)

RRXX	RWXX	RWX	WWX
(25%)	(25%)	(25%)	(25%)
Red female	Red female	Red male	White male

He rearranged the factors in symbolizing offspring, uniting Xs with each other and Rs with Ws.

At one point Morgan crossed a red-eyed male from the original wild population with a white-eyed female (yes, they were possible). Morgan's results showed roughly one-half red-eyed females, one-half white-eyed males. Morgan assumed that the white-eyed males had inherited one white-eye factor from the mother:

RX–W (red male)
WX–WX (white female)

RWXX (50%)   WWX (50%)
Red female   White male

But where had the other white-eye factor come from, if the wild flies were all red-eyed? Morgan published his conclusion: All wild males must be hybrids (heterozygous) for eye color, namely RW. Consider the implications of this hypothesis: The white-eye color allele would never express itself in a wild population, though it was present in every male. This posed a severe problem, which Morgan did not solve at first.

Describe how Morgan's conclusion was affected by his notation system. Devise an alternative system that addresses the problems it introduces. Based on your notation system, describe how the eye-color factors differ from standard Mendelian factors.

SUGGESTED READING

Allen, G. E. 1978. *Thomas Hunt Morgan: The Man and His Science.* Princeton, NJ: Princeton University Press.

Darden, L. 1992. *Theory Change in Science: Strategies from Mendelian Genetics.* Oxford, England: Oxford University Press.

Kohler, R. 1994. *Lords of the Fly.* Chicago: University of Chicago Press.

Olby, R. 1985. *Origins of Mendelism.* Chicago: University of Chicago Press.

Shine, I., and S. Wrobel. 1976. *Thomas Hunt Morgan: Pioneer of Genetics.* Lexington: University Press of Kentucky.

Wimsatt, W. 1987. "False Models as a Means to Truer Theories." In M. Nitecki and A. Hoffman, eds., *Neutral Models in Biology.* Oxford, England: Oxford University Press, pp. 23–55.

CHAPTER 6

Oswald Avery *&* the Search for the Transforming Factor

JOEL B. HAGEN

☐ *INTRODUCTION*

Today most people do not fear pneumonia, but before antibiotics became widely available in the late 1940s, the disease was a dreaded killer. In a moderately large city like Atlanta, Baltimore, or Portland (each of which had about 100,000 people in 1900), 175 to 200 victims of pneumonia might die every year. Not surprisingly, public health authorities and medical researchers placed a high priority on understanding the disease and discovering a cure for it. Oswald Avery and his laboratory at the Rockefeller Institute were at the center of this search.

Avery's work eventually led to one of the most dramatic biological discoveries of the twentieth century: the identification of DNA as the genetic material. This is surprising because Avery was not a geneticist and did not direct his research toward understanding heredity. How could the study of pneumonia open a new field of research in molecular genetics? This case study illustrates how important scientific discoveries may arise from unexpected sources. It highlights the complex relationship between pure and applied science. Although improvements in medicine often result from breakthroughs in pure scientific research, the reverse can also happen. In the case of DNA, one of the most important discoveries in modern biology grew out of the practical problem of finding a cure for a specific disease.

PNEUMOCOCCUS: THE "SUGARCOATED MICROBE"

Pneumonia is a disease with several possible causes, the most important of which is a bacterium (*Streptococcus pneumoniae*) commonly known as the pneumococcus (Figure 6.1). First isolated in 1881, the bacterium quickly became a focus of medical research. By the end of the 1920s, a great deal was known about the pneumococcus. Medical scientists knew that several strains of the bacterium existed. Antibodies that were effective against one strain had little or no effect against others. Chances of surviving pneumonia were dramatically influenced by which strain caused the infection—mortality rates ranged from 15 percent to 60 percent. At the time when Avery began his work, several major strains had been identified, each designated by a Roman numeral: types I, II, and III (today, over 80 immunological types are known).

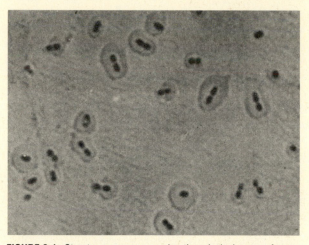

FIGURE 6.1 *Streptococcus pneumoniae*, the principal cause of pneumonia during the early decades of the twentieth century. Because of its role in causing the disease, this bacterium was often referred to as the "pneumococcus." Its early scientific name was *Diplococcus pneumoniae* because of the tendency for two spherical cells to remain attached to one another. In this photograph, the thick capsule is clearly visible around the cells. *Source:* Courtesy of the Centers for Disease Control.

Researchers also knew that disease-causing (pathogenic) pneumococci could be rendered harmless (nonpathogenic) by various artificial culturing techniques. This was usually a permanent change; harmless pneumococci produced only harmless descendants. Billions of nonpathogenic pneumococci could be injected into a mouse with no adverse effect, while only a few pathogenic bacteria quickly caused death. Types I, II, and III of the pneumococcus all included both pathogenic and nonpathogenic forms.

Scientists could easily distinguish between pathogenic and nonpathogenic pneumococci simply by looking at them. Colonies of pathogenic bacteria were larger and had a smooth, shiny appearance. Therefore, they were often referred to as the "S," or smooth, form of the bacterium. Colonies of nonpathogenic bacteria were smaller and had a rough appearance ("R," or rough, form). Under the microscope, pathogenic (S) bacteria were found to be surrounded by a thick coating, or capsule, absent in nonpathogenic (R) forms. Later, scientists discovered that the capsule was composed primarily of polysaccharides—hence Avery's facetious claim that he studied a "sugarcoated microbe." Medical researchers quickly realized that the capsule protected the bacterium against the host's defense mechanisms. Specialized white blood cells (macrophages) readily engulfed nonpathogenic (R) bacteria, but they were repelled by the capsules of pathogenic (S) forms.

The capsule also stimulated the host's immune response (see Chapter 13). Because the molecular compositions of the capsules in types IS, IIS, and IIIS pneumococci were slightly different, each caused a different antibody to be produced by the host. This claim was very controversial when Avery reported it in 1916 because most immunologists believed that only proteins could act as **antigens** (foreign molecules that stimulate antibody production). Avery had carefully purified his capsular extracts to produce a substance that was almost entirely made up of polysaccharides. This purified substance stimulated a specific immunological response when injected into a host.

The problem was that Avery could never completely eliminate minute traces of protein in his capsular extracts. Therefore, skeptics claimed that the antigenic properties could result from that residual protein. Although he turned out to be correct, Avery could not quiet his critics when he made his discovery in 1916. This technical problem of purifying bacterial extracts to eliminate all traces of protein would come back to haunt Avery in 1944 when he reported that DNA is the genetic material.

OSWALD AVERY AND THE ROCKEFELLER INSTITUTE

Avery worked at the Rockefeller Institute in New York City. Funded by oil tycoon John D. Rockefeller, this was the most prestigious medical research center in the United States. Although committed to finding cures for disease, the institute was founded on the belief that medicine must be firmly based upon science. Avery, therefore, worked closely with a diverse group of physiologists, biochemists, and biophysicists, as well as specialists in the more traditional medical fields of bacteriology and immunology.

The unique characteristics of the Rockefeller Institute provided an environment conducive to Avery's research, but his personality also contributed to his success. Former colleagues remember him most for his persistence. Once he had started a project, he rarely gave up until it was successful. The path leading to DNA was filled with obstacles that would have defeated a less tenacious investigator. "Disappointment is my daily bread. I thrive on it," he often said. This apparently fatalistic attitude was, however, based on a fundamental optimism. He took great delight in designing experiments. For Avery, the process of solving problems provided greater excitement than actually reaching the final solution. Like many other great scientists, he also inspired the younger members of his research team. This was done subtly, for he rarely assigned specific tasks to his subordinates. As he liked to claim, the Rockefeller Institute picked good, young scientists, and he "picked their brains."

EXPERIMENTS ON BACTERIAL TRANSFORMATION

In 1928, Avery and other bacteriologists were startled by Frederick Griffith's claim that he had transformed one pneumococcal type into another. Griffith was known as a meticulous scientist, but his discovery of bacterial transformation began as an accident. Before injecting experimental mice, medical researchers routinely suspended bacteria in various organic mixtures (adjuvants) in the mistaken belief that these sub-

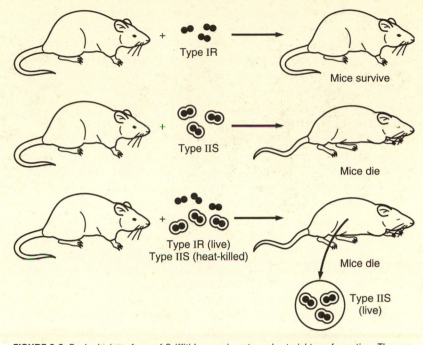

FIGURE 6.2 Basic design of one of Griffith's experiments on bacterial transformation. The use of control groups should have ruled out the possibility that some pathogenic (S) bacteria remained viable when they were heated. At first, however, Avery and other critics believed that this type of experimental error might explain transformation.

stances increased bacterial activity. Griffith just happened to use heat-killed pathogenic (type IIS) pneumococci for an adjuvant in an experiment where he injected live nonpathogenic (type IR) bacteria into mice. The mice developed pneumonia and died. When Griffith examined blood from the mice, he found live pathogenic (type IIS) pneumococci (Figure 6.2).

Somehow Griffith had transformed harmless (R) bacteria into the deadly S form simply by mixing them with an adjuvant made of dead pathogenic cells. This was surprising, but what really caught Avery's attention was the fact that Griffith had transformed pneumococci from one immunological type to another. Pathogenicity could easily be lost, so perhaps it could be gained, but most bacteriologists considered the types to be genetically distinct strains, almost equivalent to distinct species. Going from type I to type II was tantamount to converting one species into another.

Griffith's experiments caused consternation in Avery's laboratory. Avery wanted to stimulate the immune system to destroy pneumococci before they could cause disease. This "serum therapy" depended upon the stability of the immunological types of pneumococci. If types I, II, and III could change from one to another, it might be impossible to develop a cure for pneumonia.

At first, Avery hoped that Griffith's results might be due to experimental error. Griffith had done carefully controlled experiments (Figure 6.2), but perhaps he failed to kill *all* of the S pneumococci in his adjuvant. Bacteria multiply very rapidly, and mice are highly susceptible to pneumonia. Theoretically, even a single path-ogenic bacterium left in the adjuvant could quickly give rise to a lethal population. Much to his chagrin, Avery also observed transformation when he replicated Griffith's experiments. For the next 15 years, Avery and his colleagues investigated this surprising and troubling phenomenon.

Avery and Bacterial Transformation

Transformation occurred inside mice, but could it also occur in test tubes? Although Griffith attempted this experiment, he failed. Within three years of learning about transformation, Avery's lab successfully carried out *in vitro* transformation in 1931. The Rockefeller scientists mixed cultures of heat-killed pathogenic (type IIIS) pneu-mococci and live nonpathogenic (type IIR) bacteria. Later, they added antibodies specific for the type II bacteria. The antibodies destroyed any remaining type II cells that had not been transformed into type III bacteria. As a result of this selection process, all of the living bacteria harvested after incubation in the test tube were pathogenic (type IIIS) pneumococci. Although this experiment was elegantly simple in design, it was frustrating to perform. Transformation occurred only under certain laboratory conditions. Perfecting culturing techniques that consistently yielded results took months of tedious work.

> ### PROBLEM
> Avery believed that the *in vitro* experiment was the next logical step in the search for the transforming factor. Why was this step important?

Would transformation occur if Avery used an extract containing no intact cells? Numerous problems frustrated this step in the research. The original extraction process involved repeatedly freezing and thawing the culture, then heating the cul-ture for 30 minutes, centrifuging the cell fragments, and passing the resulting extract through a bacterial filter. As Avery's group discovered, this rough treatment damaged or destroyed the transforming factor. Later improvements in the technique, which involved chemically lysing, or dissolving, the cellular membrane, made it work better. It was several years, however, before Avery's laboratory could consistently cause transformation using cell-free extracts.

Ironically, during this stage in the research Avery literally had the solution to the problem in his hands. In 1932, one of his assistants precipitated a bacterial extract with alcohol. We now know that the thick, fibrous precipitate was DNA. The curious material seemed unimportant, and even if he had known what it was, Avery might have ignored it. At the time, nobody thought that DNA was an important biological molecule. Avery thought the molecule that he was searching for would almost cer-tainly turn out to be a polysaccharide or perhaps a protein.

At the time, the most likely candidate for Griffith's transforming factor was one of the polysaccharides making up the bacterial capsule. Yet one piece of evidence called this hypothesis into question. In his early experiments, Griffith found that transformation occurred when pathogenic bacteria were killed by heating to 60° C

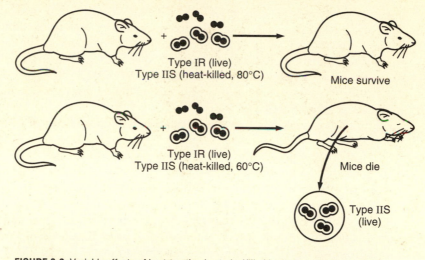

FIGURE 6.3 Variable effects of heat-treating bacteria. Killed bacteria or bacterial extracts heated to 60° C were capable of transforming nonpathogenic bacteria into pathogenic types. This ability was lost if the temperature was raised to 80° C. This experimental result seemed to rule out the possibility that the transforming factor was a polysaccharide, because most carbo-hydrates are heat stable.

but not when the temperature was raised to 80° C (Figure 6.3). This evidence sug-gested that the transforming factor was heat sensitive, but Avery knew that most polysaccharides are not damaged by such gentle heating. If not a polysaccharide, what type of molecule was the transforming factor?

Contributions of MacLeod and McCarty

Attempts to isolate and identify the transforming factor began in earnest when Colin MacLeod joined Avery's laboratory in 1934. Like Avery, MacLeod was a physician who had turned to research after a short stint of practicing medicine. He had little training in biochemical research before coming to the Rockefeller Institute—much of what he knew, he learned on the job. During the next seven years, he and Avery perfected the extraction and purification techniques that would ultimately identify the transforming factor. This was painstaking work, and much of the time MacLeod worked on his own. Avery suffered from Graves' disease, a thyroid disorder, and he was incapacitated for more than a year after having his thyroid gland removed.

MacLeod and Avery faced two types of problems: theoretical and technical. When MacLeod began working, neither he nor Avery saw transformation as a genet-ic phenomenon. After repeating Griffith's original experiments demonstrating that bacteria could change from one immunological type to another, Avery's lab adopted the following working hypothesis: Each pneumococcus contains the necessary enzymes to produce capsules of all major types (I, II, or III). Each capsule is char-acterized by a slightly different polysaccharide. Nonpathogenic bacteria of each type have lost the ability to produce any of the capsules, but this ability can be reestablished through transformation. According to the hypothesis, this process involved a stimulus mediated by the mysterious transforming factor. The type of capsule produced by a pathogenic bacterium depended on the type of transforming factor it encountered. In other words, Avery did not think of the transforming factor as a gene, which is a molecular component of the cell. Rather, he thought it was a kind of external, chemical trigger that switched enzyme systems on or off when a cell bumped into it.

This hypothesis was somewhat cumbersome because it implied that bacteria possess several enzyme systems that may never be used. Eventually, Avery and MacLeod proposed the simpler hypothesis that the transforming factor was like a gene that could be transferred from dead cells to live ones. This was not an easy step because biologists knew almost nothing about bacterial genetics. Even with this major theoretical shift, it was still necessary to chemically identify the gene. The consensus among geneticists and biochemists during the 1930s was that genes are proteins. Although scientists knew that DNA, as well as protein, could be found in the chromosomes of many higher organisms, almost everyone dismissed DNA as unimportant for heredity. One of Avery's colleagues at the Rockefeller Institute, an authority on the chemistry of nucleic acids, believed that DNA was composed of four nucleotides monotonously repeated in exactly the same order. DNA could not possibly carry genetic information, he claimed, because all DNA molecules were chemically identical. A theoretical biologist put it more bluntly: because the structure of DNA did not vary, it was "a stupid molecule."

Against this intellectual backdrop, MacLeod faced the daunting challenge of extracting enough transforming factor for chemical analysis. Although billions of bacteria can live in a liter of culture fluid, their combined mass amounts to only a few tenths of a gram. This yield was too small for the extensive chemical tests that MacLeod needed to run, but removing bacteria from larger cultures posed serious technical problems. Each liter had to be spun in a centrifuge for an hour to separate the bacteria from the fluid. At that rate it would take days of nonstop work to get enough bacteria. MacLeod turned to an alternate method: a modified cream separator borrowed from the dairy industry. He could process 50 to 75 liters of culture medium fairly quickly through the device, leaving a large cake of bacterial cells inside a 10-inch metal cylinder.

Although it provided a neat solution to the problem of harvesting bacteria, the cream separator emitted a fine mist of culture fluid. This, of course, posed an unacceptable hazard for scientists working with the bacteria responsible for causing a serious respiratory disease. Therefore, MacLeod and a technician had to modify the device to make it completely airtight. Even so, at the end of a run, they wrapped the cylinder containing the bacteria in a disinfectant-soaked towel to prevent possible contamination. The cylinder was then heated to 60° C to kill the bacteria. At this point the cake of bacteria could be safely handled. This unorthodox extraction procedure, which deviated widely from the traditional sterile techniques used by bacteriologists, bothered Avery so much that he could not bring himself to watch it.

Extraction and purification involved numerous chemical and physical techniques. The bacteria were first treated with a detergentlike substance (sodium deoxycholate) to cause the contents of the cells to be released. Various types of enzymes could be added to digest particular groups of macromolecules. The extract was then placed in a cellophane dialysis bag, which was immersed in water. Small molecules passed through the pores in the cellophane, but larger molecules remained trapped inside the bag. Various fractions of the extract could also be separated by chemical precipitation. For example, shaking with chloroform removed proteins and adding alcohol caused nucleic acids to precipitate. The physical process of centrifugation could also be used to separate parts of the extract. After these procedures were completed, the purified fractions were tested for their ability to cause transformation in living bacteria.

At the very height of this activity, MacLeod left Avery's lab. He wanted to stay at the Rockefeller Institute, but there were no permanent positions available. Therefore, when he was offered another job, MacLeod felt compelled to accept it. Although he continued to be involved in the transformation project, his new duties prevented him from playing a central role in the final discovery process. In 1941, another young physician, Maclyn McCarty, took his place in Avery's laboratory. McCarty had always wanted to be a medical researcher and, unlike his two senior collaborators, he was trained in biochemistry.

When McCarty arrived at the Rockefeller Institute, the chemical nature of the transforming factor remained unknown. Evidence for DNA had been accumulating, and one of Avery's colleagues recalls that as early as 1936 he "outlined to me that the transforming agent could hardly be a carbohydrate, did not match very well with protein, and wistfully suggested that it might be a nucleic acid!" Whatever private beliefs Avery or MacLeod had, they were unwilling to make any public statements about DNA. Indeed, they did not publish any results of their research before McCarty joined the effort. McCarty was able to build on the earlier work of the team. Combining the techniques developed by MacLeod and Avery with his own considerable chemical skills, he was able to identify DNA as the most likely candidate for the transforming factor. This happened fairly quickly. In 1943, two years after McCarty arrived, the team submitted an article on DNA and transformation to the *Journal of Experimental Medicine*, which published it the next year.

RESPONSE TO THE DISCOVERY

In contrast to the immediate acclaim that James Watson and Francis Crick would later receive for describing the structure of DNA, the response to the paper of Avery, MacLeod, and McCarty was quite muted. Relatively few readers recognized the important implication that DNA was the universal genetic material. One of the few who recognized its significance for genetics, the biochemist Erwin Chargaff, quickly switched to DNA research after reading the paper. He went on to discover that DNA nucleotides always come in specific ratios. This became a critical piece of evidence later used by Watson and Crick. Most biologists, however, remained quite skeptical about DNA.

There were several good reasons for this skepticism. Most important, Avery's group could not conclusively prove that DNA caused transformation. McCarty believed that his DNA extracts were 99.9 percent pure. But he realized that the

0.1 percent protein remaining in the sample might amount to several million molecules. Given the tremendous biological activity of many proteins, one could not rule out the possibility that it was the tiny amount of protein rather than the large quantity of DNA that was actually causing transformation.

PROBLEM
Some biologists believed that genes were composed of both protein and DNA. If McCarty's critics believed that protein was the active portion of the transforming factor, what role might DNA play? Why was the "nucleoprotein" hypothesis plausible during the early 1940s?

Avery was particularly sensitive to this possibility. In his younger years, he had faced exactly this type of criticism when he claimed that polysaccharides sometimes have antigenic properties. Much to the dismay of MacLeod, who had wanted to publish results earlier, Avery worried for months about strengthening the evidence for DNA. He did not want to be forced to retract a hasty conclusion.

In 1944, most biologists were predisposed to believe that genes were proteins. Leading biochemists believed that only proteins had the structural diversity to carry genetic information. Thus even though Avery, MacLeod, and McCarty had found that most of the transforming material was DNA, critics could still claim that the active part of the material was a protein. Avery's problem was made worse when respected colleagues, including scientists who knew much more about biochemistry than he did, criticized his work. Indeed, one of the most vocal opponents of DNA was a Rockefeller biochemist who worked in the same building with Avery. Both publicly and privately, he argued against DNA. Faced with this intense criticism by leading biochemists, it is not surprising that many biologists continued to believe that genes were either entirely protein or some combination of protein and DNA.

The problem of purification and the possible contamination of samples teaches a valuable lesson about science. Experimental results are often ambiguous and may be explained in more than one way. McCarty could never hope to attain 100 percent purity in his extracts. The acceptance of DNA as the genetic material rested upon the accumulation of evidence, only some of which was provided by Avery, MacLeod, and McCarty. For many biologists, the conclusive evidence came from later experiments, particularly the bacteriophage studies of Hershey and Chase, reported in 1952.

Other important factors also influenced the initial response to DNA. Before World War II, almost nothing was known of bacterial genetics. Bacteria were studied almost exclusively from a medical perspective. Classical genetics had developed largely around the study of *Drosophila* and other sexually reproducing organisms. There was a widespread assumption that nothing important about heredity could be learned from studying simple, asexual organisms. Even after Avery's group reported its results, one geneticist mused that although they were very interesting, it seemed unlikely that heredity in bacteria could be important because "the poor things don't have sex." As a result of this attitude, many geneticists viewed transformation as a curiosity. Ironically, much of what we now know about molecular genetics has come from studies of bacteria and viruses, but the importance of these simple organisms was not obvious to geneticists in 1944.

Finally, the time and place of the publication may have influenced the impact of the discovery. The article on DNA and transformation appeared at the height of World War II, when many scientists were engaged in war-related activities. Avery insisted that the article be published in the *Journal of Experimental Medicine*, where most of his previous work had appeared. Although this was a prestigious journal published by the Rockefeller Institute, it reached an audience made up almost exclusively of medical researchers. It was not a journal that most geneticists read. Contrast this with the rapid acceptance of Watson and Crick's article about the structure of DNA, which was published in *Nature*, one of the most widely read scientific journals in the world. Knowing your audience and publishing in the right journal are important ingredients in scientific success. Major discoveries may eventually be accepted regardless of how they are communicated, but rapid acceptance of a controversial idea may require that a large audience be quickly exposed to it.

☐ *EPILOGUE*

Oswald Avery was not looking for the genetic material when he began his research. At first, he was searching for ways to prevent or cure pneumonia. Why did his interests gradually shift away from medical research toward more general questions about the chemical and biological nature of transformation? Some historians suggest that finding a vaccine against pneumonia became less exciting with the rise of powerful antibacterial drugs (sulfanilamide and penicillin) during the 1930s and 1940s. As the threat of pneumonia receded, Avery became more interested in nonmedical aspects of transformation.

Eventually, other medical researchers did develop vaccines from capsular polysaccharides of *Streptococcus pneumoniae*, but they were administered only to patients at high risk of contracting the disease. For most people, the threat of pneumonia and other bacterial diseases became a thing of the past. Armed with antibiotic "magic bullets," the medical profession reduced mortality from pneumonia by over 95 percent after World War II. Has the tremendous success of antibiotic therapy led to complacency about disease-causing bacteria?

Although pneumococcal pneumonia is now relatively rare in the United States, other streptococcal infections pose major health problems worldwide. This is particularly true of *Streptococcus pyogenes*, which causes a wide variety of diseases including strep throat, one form of toxic shock syndrome, and several skin infections. In developing countries, 1 percent of all children suffer from rheumatic fever, a debilitating complication of strep throat. News reports of "flesh-eating" streptococci, which can quickly cause extensive tissue damage, have caused panic in both England and the United States. Although these sensational stories overstated the threat of the disease, they highlighted that serious streptococcal infections are not limited to the Third World. Knowing that antibiotic resistance has evolved in many disease-causing bacteria, reminds us that the need for effective vaccines did not end in the 1930s.

Avery's original interest in bacterial capsules still holds the key to discovering new vaccines. Unlike *S. pneumoniae*, whose pathogenicity is due to polysaccharides in the capsule, *S. pyogenes* owes its pathogenicity to a specific capsular protein. Medical scientists hope to use this "M protein" as the basis for a strep throat vaccine. Developing such a vaccine will require the techniques of molecular genetics—the field that Avery and his team pioneered half a century ago. Scientists have now cloned the gene responsible for the M protein and know a great deal about the amino acid sequence of this complex protein molecule. Although parts of the molecule vary among some 55 strains of *S. pyogenes*, other regions are constant for all strains. If the constant regions can be used as antigens, perhaps vaccines effective against all strains of *S. pyogenes* may someday be developed.

QUESTIONS AND ACTIVITIES

1. What does this case show about the following aspects of doing biology?
   — relationship between pure and applied science
   — unexpected sources of discoveries
   — role of personality in scientific discovery
   — scientific teamwork
   — burden of proof and persuasion of skeptical colleagues

2. Before World War II, almost all biologists believed that genes were proteins. Why was this a reasonable hypothesis? Why did critics believe that DNA lacked the characteristics needed for carrying genetic information?

3. The accompanying table summarizes the results from chemical analyses done by Avery's group on four purified samples of transforming factor. Do these data support the claim that DNA is the transforming factor? Do these data contradict the claim that the transforming factor is protein? How might critics of DNA have interpreted these results? How can you explain the variation among the samples and the differences between the expected and observed results?

Sample	Nitrogen (%)	Phosphorus(%)	N/P Ratio
1	14.21	8.57	1.66
2	15.93	9.09	1.75
3	15.36	9.04	1.69
4	13.40	8.45	1.58
Expected results for DNA	15.32	9.05	1.69

4. Avery, MacLeod, and McCarty discovered that adding DNA-digesting enzymes to bacterial extracts destroyed their ability to cause transformation. Protein-digesting enzymes had no effect on transformation. Do these results provide conclusive evidence that DNA is the transforming factor? How might critics of DNA have interpreted these results?

5. Does the evidence presented in Questions 3 and 4 support the claim that the transforming factor is a gene? Would this evidence force critics of Avery, MacLeod, and McCarty to conclude that DNA is the universal genetic material?

6. Study the design of Hershey and Chase's 1952 experiment (it is described in most biology textbooks). How did the results of this experiment support the hypothesis that DNA is the genetic material? To what extent would the early criticisms of DNA also apply to Hershey and Chase's experiment?

SUGGESTED READING

Amsterdamska, O. 1993. "From Pneumonia to DNA: The Research Career of Oswald T. Avery." *Historical Studies in the Physical and Biological Sciences* 24: 1–40.

Dubos, R. J. 1976. *The Professor, The Institute, and DNA*. New York: Rockefeller University Press.

Fischetti, V. A. 1991. "Streptococcal M Protein." *Scientific American* 264 (6): 58–65.

McCarty, M. 1985. *The Transforming Principle: Discovering That Genes Are Made of DNA*. New York: W. W. Norton.

Olby, R. 1975. *The Path to the Double Helix*. Seattle: University of Washington Press.

Russell, N. 1988. "Oswald Avery and the Origin of Molecular Biology." *British Journal for the History of Science* 21: 393–400.

CHAPTER 7

Hans Krebs $\mathcal{E}$ the Puzzle of Cellular Respiration

JOEL B. HAGEN

☐ INTRODUCTION

Late in life, Hans Krebs (Figure 7.1) recalled sailing past the white cliffs of Dover during the early morning hours of June 20, 1933. Although he was optimistic about a new life in England, he was experiencing a wrenching forced emigration from his native Germany. As a Jewish scientist, he had been fired from his position as a medical researcher at the University of Freiburg shortly after the Nazis came to power. Like many other German scientists, Krebs quickly fled the country. He spent several years as a refugee working in temporary research positions before becoming a British

FIGURE 7.1 Hans Krebs. *Source:* Courtesy of Philip P. Cohen.

citizen in 1939 just as World War II began. In 1953, he would be knighted and award-ed the Nobel Prize for discovering the metabolic cycle that bears his name.

Krebs was more fortunate than many other refugees. Just a year before leaving Germany, he had discovered the chemical steps by which an important waste product (urea) is produced in the liver. This was particularly significant because it was the first example of an important type of cellular process: a biochemical cycle. The discovery brought Krebs international recognition, and after fleeing Germany he was invited to work at Cambridge University, one of the world's leading centers of biochemical research.

Although the Nazis prevented him from taking more than a handful of person-al belongings, by a strange twist of fate he was allowed to take some of his labora-tory equipment. Thus his finances were in a precarious state when he arrived in England, but Krebs was able to resume his research almost immediately after settling in Cambridge. Less than five years later, he published an article announcing his most famous discovery: the citric acid cycle (Krebs cycle).

THE MAKING OF A BIOCHEMIST

Krebs grew up in Hildesheim, a town that was almost totally destroyed by Allied bombing during World War II. The son of a prominent surgeon, Krebs wanted to follow in his father's footsteps. He entered medical school shortly after the end of World War I but soon realized that he was more interested in research than in prac-ticing medicine. This was a serious decision because Germany was in the midst of a crippling economic depression and research positions were hard to find. Fortunately, after completing his medical degree, Krebs was hired to assist one of Germany's leading physiologists, Otto Warburg.

Warburg was an autocratic leader who refused to allow his assistants to work on independent projects. Despite this dictatorial style, he was an excellent teacher. He set high standards for himself and his assistants. As an outstanding physiologist who made important contributions to the study of photosynthesis, enzyme activity, and the metabolism of cancer cells, Warburg provided a role model for Krebs as the young sci-entist began his career. Krebs greatly admired his mentor, and he claimed that his own accomplishments were partly driven by a desire to emulate the great physiologist.

Warburg also perfected the instruments that Krebs used in his later research. Like many earlier physiologists, Warburg measured oxygen consumption during respiration with manometers (Figure 7.2). The Warburg manometer was unusual, however, because it could be used to study respiration in cells, rather than in whole animals. This required carefully slicing very thin pieces of animal tissue with a razor. Containing only about ten layers of cells, these slices were so thin that oxygen and carbon dioxide could freely diffuse through them. Placed in the reaction vessel of the Warburg manometer, the cells survived for several hours—plenty of time to do physiological experiments. Warburg's manometer and his "tissue slice method" provided crucial tools for discovering the details of cellular respiration. When he fled Germany in 1933, Krebs carried about 30 of these manometers with him, making his cargo far more valuable than the Nazis realized.

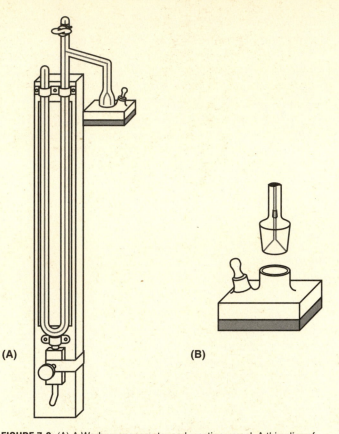

FIGURE 7.2 (A) A Warburg manometer and reaction vessel. A thin slice of tissue was placed in the reaction vessel. As respiration occurred, oxygen was consumed. The carbon dioxide produced by respiration was chemically absorbed by potassium hydroxide, placed in a small container in the reaction vessel. Because both oxygen and carbon dioxide were removed, the air pressure inside the sealed reaction vessel decreased. As a result, fluid was drawn upward into the arm of the manometer tube. Therefore, the change in the height of the fluid measured the rate of respiration. (B) Reaction vessel, showing the ground glass connection with the manometer.

THE PROBLEM OF CELLULAR RESPIRATION

When Krebs began studying respiration around 1930, scientists had already been working on the problem for more than a century. By the end of the 1700s, scientists knew that the breakdown of sugar (glucose) in living organisms was chemically similar to combustion. In both cases, a fuel is burned (oxidized) to produce carbon dioxide and water. Throughout the nineteenth century, physiologists studied this process of organic combustion in whole animals by measuring the oxygen used, the carbon dioxide produced, and the heat generated during respiration. These whole-animal studies provided useful information, but with the rise of cell theory, biologists realized that respiration must occur inside these tiny structures. Cells rather than whole organisms became the focus of later biochemical studies.

Scientists realized that respiration in cells must occur in a series of small steps, each of which releases a tiny amount of energy. Otherwise, the heat given off by burning sugar molecules would destroy the cell. There must be a kind of "biochemical pathway" leading from glucose, through a series of intermediate compounds, to the final products of respiration: carbon dioxide and water. Discovering the steps in this biochemical pathway became a major goal for twentieth-century biochemists, but they faced a serious technical problem. How can you study chemical reactions happening inside a microscopic cell?

At first, the solution to this problem seemed to be within grasp. In 1897, the German chemist Eduard Buchner demonstrated that fermentation could occur outside the cell. Using an enzyme extracted from crushed yeast cells, Buchner converted sugar to alcohol in a "cell-free" mixture. This was a dramatic scientific accomplishment, for it held out the possibility that all biochemical reactions could be duplicated in the test tube. Unfortunately, this hope was short lived. Attempts to study the oxygen-requiring steps in the breakdown of sugar using Buchner's cell-free approach failed. Unlike fermentation, aerobic respiration only occurred inside the living cell. How could scientists get inside the cell to investigate this process?

Krebs thought that he had an answer. The very thin slices of tissue that Otto Warburg placed in his manometers were really tiny colonies of cells. By experimentally manipulating the chemical environment of these cells, perhaps he could find clues to the separate reactions making up a biochemical pathway. When Krebs suggested this to Warburg, however, the older physiologist rejected the idea. At the time, Warburg was interested in comparing overall rates of respiration in normal cells and cancerous cells. Studying the separate steps in the pathway was a distraction, and Warburg refused to allow Krebs to pursue this new line of research. Studying metabolism would have to wait until Krebs left Warburg's laboratory.

BIOCHEMICAL PUZZLE SOLVING

A few years later, Krebs moved to another laboratory at the University of Freiburg, and he began seriously studying cellular respiration in 1932, shortly before he fled Germany. At the time relatively little was known about the process. Following the lead of Eduard Buchner, scientists realized that something similar to fermentation occurred in all cells. A molecule of glucose was always split into two pieces (pyruvate) during the first stage of the breakdown of sugar. This set of reactions (what we now call *glycolysis*) did not require oxygen. What happened to these glucose fragments during the oxygen-consuming steps of cellular respiration still remained a mystery, although many scientists were working on the problem.

Krebs later claimed that solving the problem of cellular respiration was like putting together a jigsaw puzzle. Brainstorming with pencil and paper, he generated numerous hypothetical pathways. Each chemical step in a pathway had to follow the rules of organic chemistry, but just because it was plausible did not make it correct.

In fact, most of the hypotheses turned out to be wrong. The only way Krebs knew whether a hypothetical step could actually be a piece in the puzzle was to test it with the Warburg manometer. Suppose that a hypothetical chemical compound really was one of the intermediate steps in the chain leading from glucose to carbon dioxide. If so, then adding this substance to the tissue slice in the Warburg manometer ought to increase the rate of oxygen consumption.

The Search for Intermediate Compounds

To visualize this type of experiment, remember that respiration is somewhat similar to combustion. Adding fuel to a fire increases combustion, so more oxygen gets consumed. Following this analogy, glucose is fresh wood, and the intermediate compounds are partially burned embers. Compounds unrelated to cellular respiration are not fuel and so do not burn at all. During the early 1930s, Krebs and other biochemists tested many potential "fuels" to find out whether they would "burn." For example, in one set of experiments Krebs tested three 4-carbon compounds: succinate, fumarate, and malate. Could they be intermediate compounds in the respiratory pathway? Because all three compounds increased oxygen consumption, it seemed likely that they were involved in cellular respiration. Using the combustion analogy, Krebs was adding more embers to the fire, so more oxygen was being used to burn them. When he completed the experiments he discovered the following results:

Substance Added	Oxygen Consumed (μmole)
None (control)	670
Succinate	1,520
Fumarate	1,290
Malate	1,340

Like many experiments, however, these raised more questions than they answered. Succinate, fumarate, and malate increased respiration even more than Krebs expected. Furthermore, all three of the compounds could still be detected in the reaction vessel at the end of the experiment. If they really were intermediate compounds in the slow oxidation of glucose to carbon dioxide, why weren't these extra embers completely burned in the process? If they weren't intermediate compounds, why did respiration increase so much when they were added to the reaction vessel?

The Hungarian biochemist Albert Szent-Györgyi, who later won a Nobel Prize for discovering Vitamin C, proposed one plausible explanation for the unexpected results. He believed that succinate, fumarate, and malate were not reactants in cellular respiration. They acted like catalysts, instead, accelerating the chemical reactions of cellular respiration without being consumed in the process. This explained why they were still detected at the end of the experiment. If this were true, however, biochemists would have to keep looking for the true reactants. Were the three compounds catalysts or reactants? The controversy remained unresolved for almost two years.

A Role for Citric Acid

Many other pieces of the jigsaw puzzle were also missing. Perhaps the most puzzling was citric acid (citrate), a six-carbon molecule. Biochemists could increase cellular respiration by adding citrate to the Warburg manometer, but they were unable to find a logical place for citrate in their hypothetical pathways. Krebs's early attempts to brainstorm a role for citrate had failed, and he never published them.

> *PROBLEM*
> **Both glucose and citrate are six-carbon molecules. Knowing what they did about glycolysis, why might it have been difficult for biochemists to hypothesize a role for citrate in the later stages of cellular respiration?**

In 1937, two German biochemists, Franz Knoop and Carl Martius, published a series of reactions beginning with citrate and ending in another organic acid, oxaloacetate (see Figure 7.3(A)). Prominently placed in this scheme were the three compounds

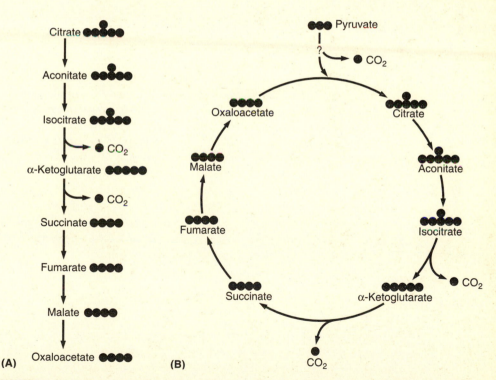

FIGURE 7.3 (A) The hypothetical metabolic pathway proposed by Franz Knoop and Carl Martius. This pathway included all of the steps in the Krebs cycle in the correct order. But Knoop and Martius incorrectly presented the reactions as a linear pathway rather than a metabolic cycle. *Note:* Only the carbon backbones of each intermediate compound are represented in this diagram. (B) A simplified diagram of the Krebs cycle. The crucial step leading from pyruvate into the cycle was not fully explained for a decade after the publication of Krebs's scheme. The linkage between the steps in the Krebs cycle and the production of ATP was also not well understood in 1937.

studied by Krebs and Szent-Györgyi: succinate, fumarate, and malate. Was this a large missing piece in the biochemical pathway leading from glucose to carbon dioxide?

After reading this report, Krebs designed a simple experiment to test this hypothesis. He would try to block one of the reactions in the hypothetical pathway with a chemical inhibitor. If successful, the later reactions in the chain should not occur and the intermediate compound just before the block should accumulate. To visualize the design of this experiment, think about a mountain stream cascading from one pool to another as it flows downstream. If you dam the stream, the water will stop flowing below the dam and will accumulate in the pool just above the dam.

Krebs chose to inhibit the proposed reaction leading from succinate to fumarate (see Figure 7.3(A)). If the Knoop-Martius pathway was correct, succinate should accumulate in the Warburg manometer. At the end of the experiment, Krebs recorded the following results:

Substance Added	Succinate Produced (µl)
None (control)	20.5
Citrate	64.5
Citrate plus inhibitor	387.0

PROBLEM

Think about the mountain stream analogy. Why did succinate accumulate when citrate was added to the Warburg manometer? Why did even more succinate accumulate when both citrate and the inhibitor were added? Suppose that Krebs had measured the amount of oxaloacetate produced during this experiment. What results would you expect if the Knoop-Martius pathway were correct? *Note:* Refer to Figure 7.3(A) when answering these questions.

Tests with other inhibitors produced similar results. These experiments seemed to support the hypothesis of Knoop and Martius, but there was still the nagging question about the fate of succinate, fumarate, and malate. Why didn't they disappear during respiration? According to the Knoop-Martius hypothesis, they should be burned up during the slow combustion of glucose.

Pathway or Cycle?

At this point, Krebs began to think in terms of a biochemical cycle rather than a linear pathway. This was a possibility that no one, including Krebs, had previously considered (see Figure 7.3(B)). If the ends of the Knoop-Martius pathway were connected, several pieces of the jigsaw puzzle would fall into place. A cycle of reactions would explain why succinate, fumarate, and malate did not disappear during cellular respiration—instead of being completely oxidized, they were regenerated with each turn of the cycle. It also explained the crucial role of citric acid as the first intermediate compound in the cycle. Finally, the results of Krebs's experiments with chemical inhibitors fit just as well with a biochemical cycle as they did with a linear pathway.

Could Krebs find experimental evidence that would support his hypothesis rather than the hypothesis of Knoop and Martius? The key would be to demonstrate a reaction that produced citrate from oxaloacetate. In a crucial set of experiments, Krebs found that when both pyruvate and oxaloacetate were added to the Warburg manometer, the amount of citrate increased. This was particularly significant because Krebs knew that pyruvate was the major product of glycolysis. He had firmly tied the two ends of the Knoop-Martius pathway together, forming what we now call the Krebs cycle.

At this point—four months after reading the article by Knoop and Martius—Krebs was ready to report what he called "the citric acid cycle." He chose to submit a short article to *Nature*, one of the world's most widely read scientific journals. Because it is published weekly, scientists often use this journal to quickly communicate important discoveries. Ironically, *Nature* rejected the paper—the only time this ever happened to Krebs. Perhaps we should not make too much of this, because almost every scientist has papers rejected. The editor was not a biochemist, and he had to select among many submitted manuscripts. This anecdote should remind us, however, that sometimes even important scientific discoveries are not immediately recognized.

Stung by the rejection, Krebs sent a longer article to a leading biochemical journal, which quickly published it. Realizing the importance of Krebs's work, biochemists expanded on it. During the years following its discovery, the citric acid cycle was found in a wide variety of organisms, including many bacteria. It played a critical role in the oxidation not only of glucose and other carbohydrates, but also of fats and proteins. What later became known as the "Krebs cycle" was the central energy-releasing pathway in almost all cells.

Krebs liked to compare his accomplishment to piecing together a jigsaw puzzle. As he recognized, however, the Krebs cycle was also part of a much larger puzzle. During his research, Krebs treated the cell as a "black box." Neither he nor anyone else knew exactly where respiration occurred in the cell. Although some scientists suspected it occurred in the mitochondria, this was not demonstrated until a decade after Krebs's discovery. The important role of ATP as the primary energy-carrying molecule was only beginning to be understood when Krebs discovered his cycle. How ATP is synthesized in the mitochondria would not be discovered for another 30 years (see Chapter 8). Although Krebs knew that pyruvate links glycolysis to the citric acid cycle, he did not understand the connection. The important role of coenzyme A in bridging these two pathways was not discovered until a decade after Krebs described his cycle. Recognizing how many of the pieces of the puzzle were still missing in 1937 does not diminish Krebs's accomplishment, but it reminds us that scientific knowledge is often constructed of small increments.

☐ *EPILOGUE*

How can we explain scientific creativity? What makes one scientist more successful than another? Why was Krebs able to see a biochemical cycle where other scientists still saw a linear pathway? There are no simple answers to these questions, but Krebs's scientific career provides some interesting insights into the nature of creativity. It also dispels a common misconception about how discoveries are made.

Remember that Krebs had discovered another biochemical cycle several years before he completed his work on the citric acid cycle. The urea-producing process was the first example of a biochemical cycle, and its discovery brought Krebs international recognition. Did this early discovery predispose Krebs to look for other biochemical cycles? Is there evidence that at some point in his later research on cellular respiration he had a flash of insight in which he suddenly recognized another cycle? Apparently this kind of "Aha!" or "Eureka!" experience did not occur. Krebs could never pinpoint exactly when the discovery took place, and he always denied having had an "Aha!" experience. Frederic Holmes, a historian who carefully studied Krebs's laboratory notebooks, also claims that there is no evidence for a sudden flash of insight. Like other biochemists, Krebs believed that cellular respiration was a linear pathway, and he only gradually came to see its cyclic nature.

If Krebs's discovery was not due to a flash of insight, what might account for his achievement? Holmes points to several important elements of Krebs's creativity. Like all successful scientists, Krebs was deeply committed to research. Early in his career, he pinpointed the major unsolved problems facing the biochemists of his day. After carefully choosing among these problems, he devoted his life to solving some of them. This usually meant spending six days a week in the laboratory. Despite this rigorous, highly structured approach to work, Krebs remained intellectually very flexible. He quickly abandoned hypotheses when they could not be supported by experimental evidence. This willingness to change direction meant that he rarely got bogged down in fruitless dead ends.

These characteristics help explain Krebs's creative insight. He had been trying to fit the pieces of his puzzle together for several years. Although he did not have a solution to the complete problem, he was very familiar with many of its parts. When he ultimately solved the problem, he was following the same routines that he had used throughout his research. There was no flash of insight, just a reworking of the problem from a slightly different perspective. This fresh perspective emerged after reading the paper of Knoop and Martius. Even so, it took several days of thinking and experimenting before Krebs began to realize that tying the ends of the Knoop-Martius pathway would produce another biochemical cycle. Gradually seeing what no one had previously recognized was Krebs's creative act.

QUESTIONS AND ACTIVITIES

1. What does this case show about the following aspects of doing biology?
   — role of scientific instruments
   — mentors and the training of young scientists
   — positive role of incorrect hypotheses
   — incremental nature of discovery
   — scientific creativity

2. During the 1800s, Eduard Buchner performed experiments using "cell-free" extracts. He demonstrated that fermentation occurred even in the absence of intact cells, but he was unable to detect respiration in cell-free extracts. How can this be explained by the different sites where these processes occur in the cell? Why did respiration occur in tissue preparations that Warburg placed in his manometer?

3. When Krebs added succinate, fumarate, or malate to the Warburg manometer, he was surprised to find that respiration increased even more than predicted. He was also surprised to find that these compounds were not consumed during respiration. How does the Krebs cycle—Figure 7.3(B)—explain these unexpected experimental results? What would you predict would happen if pyruvate were added to the Warburg manometer? Why might pyruvate behave differently than succinate, fumarate, or malate?

4. Krebs compared doing science to putting together a jigsaw puzzle. He stressed the importance of cumulative small steps in problem solving. How did the type of questions Krebs asked, his research style, and his equipment favor a puzzle-solving approach to scientific work?

5. Consider the Krebs cycle as it is presented in your biology textbook. What parts of the process were known before Krebs published his description of the cycle? What contributions did Krebs make? What important parts were discovered after Krebs published his paper?

6. Krebs called his cycle the "tricarboxylic acid cycle" after the family of chemical compounds involved in the reactions. He also used the term "citric acid cycle," after the first intermediate in the cycle. Today most biologists refer to this process as "the Krebs cycle." Does this honorific name give too much credit to one scientist among the many who worked on the problem of cellular respiration? How much credit for the discovery do Knoop and Martius deserve for proposing their hypothesis?

SUGGESTED READING

Allen, G. 1975. *Life Science in the Twentieth Century.* New York: John Wiley.

Fruton, J. S. 1972. *Molecules and Life: Historical Essays on the Interplay of Chemistry and Biology.* New York: Wiley-Interscience.

Holmes, F. L. 1991. *Hans Krebs: The Formation of a Scientific Life 1900–1933.* Vol. 1. New York: Oxford University Press.

Holmes, F. L. 1992. "Manometers, Tissue Slices, and Intermediary Metabolism." In A. E. Clarke and J. H. Fujimura, eds., *The Right Tools for the Job: At Work in Twentieth-Century Life Sciences.* Princeton, NJ: Princeton University Press.

Holmes, F. L. 1993. *Hans Krebs: Architect of Intermediary Metabolism: 1933–1937.* Vol. 2. New York: Oxford University Press.

Krebs, H. A. 1970. "The History of the Tricarboxylic Acid Cycle." *Perspectives in Biology and Medicine* 14: 154–170.

Krebs, H. A. 1981. *Reminiscences and Reflections.* Oxford, England: Clarendon Press.

CHAPTER 8

Peter Mitchell & How Cells Make ATP

DOUGLAS ALLCHIN

☐ INTRODUCTION

Science sometimes exudes an aura of the solidity of its facts. In 1963, the study of energy transformations in the cell was far from this image, however. One researcher, at least, could joke about it, observing that "Anyone who is not thoroughly confused just does not understand the situation."

Yet the focus of research—the process whereby cells convert energy from the Krebs cycle into a usable form—was central to the cell. This was where the oxygen we breathe is ultimately used. It was where cyanide, arsenic, and several pesticides can have their deadly effects. It was where cells make most of their **adenosine triphosphate**, or **ATP**—the primary energy-carrying molecule. We use ATP in our muscle cells, for example, to walk, to grasp, to speak, to swallow, to blink, to inhale, to pump blood. Cells depend on the reactions that produce ATP for various other chemical functions as well. Indeed, to the extent that energy is essential to maintaining the organization of living matter, the problem was akin to asking, "What is life?" The reactions were vital. Yet no one knew how cells make ATP.

Biochemists tried to trace the energetic pathways, applying many of the same strategies used by Hans Krebs and his colleagues (Chapter 7), but without success. They had evidence that there were several intermediate steps, and they had set about to isolate and identify the intermediate molecules. Over many years, different chemists announced that they had succeeded, but each claim later turned out to be erroneous. The series of apparent successes, each mistaken, was disheartening. One text writer noted the irony for his student readers: "No worse fate could befall anyone working on [the problem] than to solve it." Failures in science may often recede gracefully into obscurity, but on this occasion another prominent researcher admitted that the chemists' efforts had met with "conspicuous non-success."

Such a field would seem ripe for new ideas. And in fact, it was against this backdrop that Peter Mitchell (Figure 8.1) introduced a novel hypothesis. It was a radical and, in retrospect, triumphant new solution. Mitchell eventually received a Nobel

FIGURE 8.1 Peter Mitchell (left) with his colleague, Jennifer Moyle. *Source:* Courtesy of Glynn Laboratories, Bodmin, England.

Prize for his insights. Yet his ideas were not widely adopted for almost a decade and a half, despite the chemists' continuing frustrations. How did Mitchell solve the vexing puzzle? And why did other chemists not immediately embrace his solution?

A PREHISTORY OF CHEMIOSMOTIC IDEAS

Peter Mitchell's ideas were revolutionary, leading many to rethink chemistry itself in biological contexts. But for Mitchell, the ideas hardly seemed revolutionary at all. They emerged naturally from other ideas in his experience. They had roots extending back many years, to before he even addressed the problem of how cells make ATP. In this case, the history of Mitchell's thinking can help someone understand the ideas themselves.

While a graduate student in biochemistry at Cambridge University in the early 1940s, Peter Mitchell worked with cell membranes. Membranes pose interesting problems because they are barriers that help maintain the vital integrity of cells. Certain molecules enter cells, while others do not. As a result, the fluid inside a membrane is different from the fluid outside. A membrane thus functions as an **osmotic barrier**. Mitchell worked closely with a chemist who asked, how do such barriers work?

Mitchell became fascinated with one aspect of the problem: How are certain materials "actively" transported across the membranes of cells? Some molecules pass through membranes easily. They diffuse passively as though through a filter. Other molecules, by contrast, need energy to actively cross the osmotic barrier. This is especially true for molecules like nutrients that must concentrate inside the cell. Molecules do not tend to concentrate on their own—quite the opposite: they tend to disperse. Concentrating them (by actively transporting them inside a membrane) requires energy. Mitchell was curious: how is energy involved in moving the molecules?

Scientists often use analogies to think about an unfamiliar problem. Here, you might imagine familiar examples where energy causes movement. For instance, water releases energy as it falls, and the force of its fall can turn a waterwheel. Likewise, a compressed spring can release energy and propel the movements of a toy vehicle. Could something perhaps "fall" or "spring open" on a cellular level in a way that might help move molecules across a membrane?

PROBLEM
Describe a few examples, from your everyday experience, where energy in various forms causes movement. (Consider, for example, batteries, engines, pressurized air or steam, hand pumping, gravity, etc.) Where possible, describe how each example suggests an image for thinking about how molecules might move across a cellular membrane.

In the late 1940s, two largely independent groups of chemists each saw the problem of membrane transport differently. One group of chemists studied membranes. They looked at the physical structure of membranes and how they allowed certain substances to pass through. A second group of chemists studied proteins, enzymes, and the reactions they catalyzed. They recognized the role of energy in Mitchell's problem, but their models of how enzymes worked did not include movement. Mitchell was unusual in thinking about both branches of cellular chemistry at the same time. He thus saw transport as both a physical phenomenon *and* a chemical reaction using energy. The chemical reaction changed reactants on one side of the membrane into products on the other side of the membrane. "Active" transport was a chemical reaction that spanned the membrane.

Later, Mitchell coined a new term, **chemiosmotic**, to describe these membrane-spanning types of reactions: '*chemi*' because the reactions were chemical reactions, '*osmotic*' because they occurred across an osmotic barrier, or membrane. Mitchell drew on the Greek root, '*osmo*' meaning "to push." The reactions that Mitchell described were not at all related to the physical process of osmosis, though the words share a common root. In fact, Mitchell was upset when textbooks later introduced the term '*chemiosmosis*' a label he felt was grossly misleading. But how could he prevent others from misrepresenting his ideas?

In the early years of his career, Mitchell worked both on his own and in collaboration with Jennifer Moyle, a colleague from Cambridge (Figure 8.1). He experimented on how bacteria transported various compounds across their membranes. At the same time, he reflected theoretically on his problem. Chemical reactions, Mitchell had concluded, occur in three-dimensional space. But how could enzymes catalyze such reactions? Mitchell explored his ideas concretely by building a series of wooden mechanical models of membrane enzymes (Figure 8.2). The model enzymes would first take two reactants from one side of a membrane into a central channel. When they reacted, the enzymes would use energy to change shape. The entering channels would be closed off, and a new exit channel would open on the other side. They

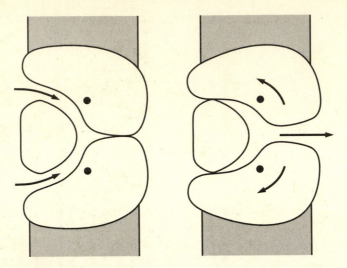

FIGURE 8.2 "Before" and "after" views of a membrane protein, according to Mitchell's early ideas.

would work like one-way turnstiles or the triggered gates found in subways and sports stadiums. Mitchell shared each model with colleagues, asked their opinions, and then revised his models. The models were a way to visualize and develop the notion that transport reactions had direction.

At one point, Mitchell saw the relationship between energy and transport in reverse. Whereas before he had thought only about how energy was used to transport molecules, he now began to think about how the movement of molecules might be used to fuel energy-requiring reactions in the cell. In other words, a molecule traveling from one side of a membrane (where its concentration was high), to the other side (where its concentration was low) would release energy. Could that energy be used to make ATP? To think about the problem, you would reverse the analogies introduced earlier. For example, for a spring to have the *energy* to run a toy, someone has to compress the spring—a form of *movement*. Similarly, *turning* a waterwheel can raise water—*working* against gravity.

PROBLEM
Consider again your earlier analogies linking energy and movement. Describe those cases where the movement could generate energy that could be used or stored. If no case applies, identify another case where movement provides the energy for another process, or where the movement stores energy in a form that can be used later.

In the late 1950s, Mitchell capitalized on a finding from one of his graduate students. She had found that the transport of a particular sugar in bacteria was coupled with the movement of another reactant: a hydrogen ion, or proton. Mitchell recognized that the particles that moved across membranes might be as simple as protons. If the concentration of protons was higher on one side of the membrane, a cell might be able to make ATP when the protons moved.

Mitchell had been working with bacteria, but now he turned his thoughts to cases of ATP production in more complex cells. In eukaryotic cells ATP is produced in the membrane of one organelle, the mitochondrion. ATP is also produced in similar ways in the membranes of chloroplasts in plant cells. Might protons be involved in making ATP there? What would create an imbalance of protons to begin with? Mitchell had ideas about this, as well, and he dispatched a short, speculative article that was published in 1961.

Not long thereafter, Mitchell's professional career took a sharp detour for personal reasons. While teaching at the University of Edinburgh, Mitchell developed gastric ulcers. The institutional environment, he noted later, "did not suit his temperament." At age 43, he was forced to retire from formal academic life. Mitchell moved to the rural region of Cornwall in southern England. There, according to his own testimony, the hand-milking of eight Jersey cows morning and evening for several months did wonders for his ulcers. At the same time, Mitchell nurtured a vision of a small, private research institute where he could work in a less bureaucratic setting. He invited his longtime collaborator, Jennifer Moyle, to contribute her strong experimental skills, and she agreed to join the venture. Mitchell's brother offered the funds. Over the next two years, Mitchell and his family helped renovate a crumbling nineteenth-century mansion into a complex of labs, offices, a library, and a suite of rooms that would be his home. During this three-year pause in research, Mitchell published occasional theoretical papers. But with no lab, his ideas remained relatively dormant.

In 1965, Mitchell was invited to present his chemiosmotic ideas at a conference. Mitchell wanted to present some original data, so he and Moyle hastily assembled some inexpensive and partially homemade equipment. They were able to measure changes in proton concentration (pH) in the solution outside the mitochondria, as predicted by Mitchell's emerging theory. In fact, the results were much better than they expected. Mitchell reported their results at the conference, but few there accepted his ideas. Indeed, a controversy over how cells produce ATP soon flared.

INTERSECTING RESEARCH TRADITIONS

Mitchell had proposed his new ideas in a field that was already well established and had a flourishing tradition of research. After Krebs, chemists were still puzzled about how most of the energy from the citric acid cycle made ATP. The cycle channeled energy into a particular form: high-energy electrons attached to electron-carrier molecules known as NAD and FAD. By the early 1950s, chemists had traced the pathway of energy further. They knew that NAD and FAD transfer their high-energy electrons to a series of molecules in the inner mitochondrial membrane known as the electron transport chain (or the respiratory chain or oxidation chain). In the electron transport chain, the electrons shift to successively lower energy states as they move from molecule to molecule. At the same time, energy is released. The energy that is released produces ATP.

Electrons eventually reach their lowest level when they combine with oxygen and protons, making water. Here is the cellular process associated with our need to breathe. By accepting electrons at the end of the chain, oxygen allows electrons to flow and to transfer their stored energy to ATP. When oxygen is absent, electrons stop flowing and no ATP is made. In the heart, for example, lack of oxygen means that no ATP is made for the heart muscle cells to contract; as a result, the person has a heart attack. Chemists after Krebs turned to investigate this central question: "How is ATP made as a result of electron transport?"

Chemists were accustomed to thinking in terms of reaction pathways. They reasoned that there must be more intermediate steps. More steps implied more intermediate molecules holding the energy and more enzymes to catalyze each reaction—making what was, for them, a complex process even more complex! Chemists measured the drops in energy levels of electrons to find where enough energy would be released to produce ATP molecules. As Krebs and others before him had done, they set out to find the intermediate molecules.

Mitchell approached the problem of ATP production from an entirely different perspective. The resulting clash highlights, first, just how much the processes Mitchell described differ from other energy reactions in the cell (such as glycolysis and the Krebs cycle). Second, the episode illustrates how scientific knowledge can sometimes undergo revolutionary shifts in perspective.

The electron transport chain was embedded in the mitochondrial membrane and so, Mitchell imagined, the energy system might involve transport across that membrane. In particular, the movement of negatively charged electrons between electron carriers might, through electrical interaction, move positively charged protons across the membrane. In this view, there would be no intermediate molecules as chemists had assumed. Rather, the intermediate energy state would be a different kind of energy: an imbalance of protons on either side of the mitochondrial membrane. The electron transport chain would first create the imbalance. Then the protons would move to restore the balance by traveling through an enzyme, making ATP as suggested in Mitchell's earlier turnstilelike models (Figure 8.2).

For Mitchell, the critical element was how electrons, in cascading down energy levels, could move protons across the membrane. His scheme was based on the idea that as electrons drop energy levels, they also move through space. The electron would start at the first carrier molecule in the electron transport chain, positioned on one side of the membrane. Because of electrical attraction, the negatively charged electron would attract a positively charged proton from the water on that side of the membrane (Figure 8.3(A)). Then the electron would naturally drop energy levels, but because the next electron-carrier molecule was positioned on the other side of the membrane, the electron—with the proton "in tow"—would move across the membrane (Figure 8.3(B)). It would be as if you were riding a pulley down an inclined rope across a river. You would move because gravity pulled you down, but the movement would also take you across the river because the rope would guide how you "fell." If someone grabbed onto you (just as the proton was attracted to the electron), he or she could be carried across the river along with you. The proton, as a "free rider," would now be on the other side of the membrane.

When the electron combined with the electron-carrier molecule on the other side of the membrane, Mitchell postulated, it would no longer be able to hold on to the proton. The proton would then be "dumped" (Figure 8.3(C)). Merely by having

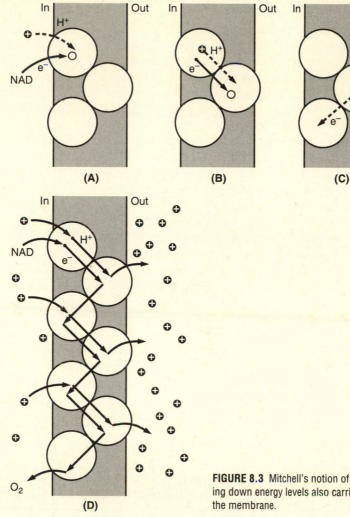

FIGURE 8.3 Mitchell's notion of how electrons flowing down energy levels also carried protons across the membrane.

been attracted to the electron, the proton would have been ferried to the other side of the membrane.

The process would then be repeated as the electron continued to drop energy levels from one electron-carrier molecule to another and onto oxygen. The net effect would be to move many protons from one side of the membrane to the other, thus creating a reservoir or imbalance of protons outside (Figure 8.3(D)). As noted earlier, the high concentration of protons would fuel an enzyme that made ATP as the final step. For Mitchell, the whole process would work because the electron carriers were positioned alternately on opposite sides of the membrane. Here, the chemistry of living things depended on their unique molecular organization.

Mitchell's view of ATP production highlighted several puzzles in the ongoing research on mitochondria. For example, no one had been able to reproduce the mitochondrial reactions in a test tube without an intact membrane being present. Indeed, this had been a nuisance for chemists trying to isolate and study individual components of the electron transport chain. For Mitchell, the mitochondrial membrane kept protons on one side of the membrane once they had been "ferried" across. If the membrane ever broke, the protons would dissipate and even out again; the energy would be lost and no ATP could be made. No wonder a membrane seemed essential.

Other chemists interpreted the problem of the membrane differently, however. They thought that the membrane might act like a scaffolding or skeleton, keeping the various parts of the electron transport chain positioned so they could interact properly. The membrane would be essential, but for a very different reason than the one that Mitchell proposed. Mitchell was able to explain the need for a membrane, but Mitchell's ideas were not the only plausible explanation.

To researchers already working on how cells make ATP, Mitchell's proposals about protons and membranes were undisciplined speculation. They saw Mitchell as someone unfamiliar with the field and inadequately trained in energy reactions. Further, Mitchell had studied bacteria, not the mitochondria of the more complex eukaryotic cells. The theory of the endosymbiotic origin of mitochondria (Chapter 3) had not yet been widely recognized, and few chemists perceived studies of energy processing in bacteria as relevant to mitochondria (see also Chapter 6 for a similar earlier bias in genetics). One chemist summarized the prevailing attitude: "I do not believe that the only or even the primary function of this incredibly complicated chain of reactions is to produce protons at the right place." And so debate continued through the 1960s.

STRATEGIES FOR RESOLVING DISAGREEMENT

How did chemists ever come to accept the chemiosmotic hypothesis? The controversy was certainly not resolved overnight. First, Mitchell did not retreat from his ideas when they were severely criticized. Though Mitchell's disposition was generally gentle and friendly, he was quite tenacious when it came to his ideas. In fact, according to the younger generation of chemists, *all* the leading chemists working during this period had strong egos. There was no doubt, given the magnitude and centrality of the problem, that a Nobel Prize hung in the balance. And the major researchers, including Mitchell, wanted it.

Second, the controversy was resolved in part because the ideas themselves changed. The specific concepts that Mitchell proposed originally and those that chemists came to accept later differed significantly. Mitchell repeatedly revised his theory to address both conceptual problems and new evidence. In his first article, for example, Mitchell had protons moving in the wrong direction! It was a simple mistake, but clearly important for a chemist trying to measure the process. Also, Mitchell had largely guessed about the number of protons needed to make each ATP molecule, and others ridiculed his tentative suggestions as wildly unrealistic. In addition, the proton imbalance that Mitchell postulated was unprecedented and would likely destroy cells. Mitchell's original notion of the alternating position of electron carriers also did not fit the actual data. He eventually abandoned this concept, replacing it with a far more sophisticated version that other researchers agreed explained the available data very nicely.

Mitchell was both persistent and flexible. He was unwilling to sacrifice his guiding notion that reactions occur in three dimensions, across membranes—and he worked hard to find arrangements that fit the data. Given the limited resources at his private research institute, he had to focus his efforts. In the afternoon, as he sipped tea, he mulled over theoretical problems and identified an experiment to provide critical information. The following morning Moyle would perform the experiment (Mitchell was never known for having good lab skills!). In the afternoon, they would discuss the theoretical consequences of the results. Mitchell would mull some more, sometimes consulting sources in his library, and the two of them would plan where to head next. The close interplay between Mitchell's theoretical adjustments and Moyle's experimental results was essential in revising chemiosmotic interpretations to address everyone's concerns.

Third, Mitchell worked personally with other chemists. Located in a small lab out of the mainstream, Mitchell might easily have become isolated from the scientific community. But he maintained active correspondence with other leading thinkers in the field. He telephoned others when he had specific disagreements with them, presented his point of view, and tried to resolve differences. He also invited others—including his harshest critics—to his lab. Over a one- to three-day visit, they discussed ideas and often conducted short experiments together. These personal visits did not always convince the visiting researchers immediately, but they did help them appreciate the chemiosmotic perspective. And given Mitchell's radical departure from earlier ideas in the field, that sense of understanding was valuable.

AN ARTIFICIAL REALITY

Sometimes, researchers can construct a single crucial experiment to help them decide between two sharply contrasting theories. But here the ideas were too complex. Instead, many tests, each addressing a separate question, were needed to validate Mitchell's ideas. One experiment is especially noteworthy, however, because it exemplifies the kind of dramatic, well-designed experiment that most scientists dream of performing. It also shows how even important results can often be interpreted in different ways.

As Mitchell was completing the renovation of the mansion for his research institute, he received a letter from André Jagendorf at Cornell University. Jagendorf reported that he had encountered some observations in plant cells that confirmed aspects of Mitchell's 1961 hypothesis. He and a graduate student had measured the proton changes due to electron transport in chloroplasts. (These reactions, associated with chlorophyll's use of light energy, closely parallel those in mitochondria.) That was certainly promising. But how could others be sure that the changes were not a coincidence, caused by some unknown factor?

Jagendorf's strategy was to *create* an imbalance of protons artificially and document its effects. First, Jagendorf and a colleague incubated chloroplasts in an acid bath, allowing protons to saturate the solution both inside and outside the chloroplast membrane. They then plunged the chloroplasts into a solution with a lower concentration of protons. This created an imbalance of protons across the membrane. ATP was produced! Because they had conducted the experiment in the dark, they could be confident that the chloroplasts had not used light energy. Rather, the induced imbalance in protons must have made the ATP, mimicking the situation in real chloroplasts, where light was present.

The acid-bath studies were dramatic. The experimenters had induced chloroplasts to perform a natural function under artificial conditions. They had demonstrated that nature behaved in ways that biologists and biochemists had not previously suspected. And they had done so by creating an "unnatural" situation. For many who were accustomed to thinking about intermediate molecules, the experiment was breathtaking.

By today's reckoning, the acid-bath experiments seem like unassailable proof of the chemiosmotic hypothesis (now chemiosmotic "theory"). And indeed, they did confirm one of its controversial predictions. But they did not prove the theory. They did, nonetheless, *demonstrate* an impressive, novel finding. Before the demonstration, the question was whether an imbalance of protons could generate ATP at all. After the experiments, the question was whether ATP could be produced *without* protons. The burden of proof had shifted. Jagendorf's results did not *disprove* the ideas about chemical intermediates (see Question 5 at the end of the chapter). But they did significantly alter the horizon of debate. That was the power of the demonstration. And it was through a series of such demonstrations that chemists began to pursue chemiosmotic ideas, without ever quite explicitly renouncing their earlier ideas.

☐ *EPILOGUE*

Mitchell received a Nobel Prize in 1978. Later he was portrayed somewhat playfully in a cartoon (Figure 8.4) as the Christopher Columbus of bioenergetics (the field that studies energy in the cell). The caption tells us that "Mitchell sets sail for the Chemiosmotic New World, despite dire warnings that he will be consumed." Note the naysayer on the dock, whose emphatic squiggle, ~, denotes the energy in the hypothesized intermediate molecules. What does this image convey about the process of science and the role of individual genius? (See Question 2.) Mitchell once remarked that "science is not a game like golf, played in solitude." Rather, he observed, it is "a game like tennis in which one sends the ball into the opposing court, and," he added, "expects its return."

Disagreement and controversy can sometimes be construed as symptomatic of weakness in science. If scientists cannot decide between alternative theories, then it may seem as if they do not have strong evidence for either theory. In most cases, however, disputes arise because scientists know too much. In this episode, strong evidence supported each view, yet there seemed no way to reconcile the sometimes conflicting conclusions. Scientists frequently debate theories. This typically means

FIGURE 8.4 "Mitchell sets sail for the Chemiosmotic New World, despite dire warnings that he will be consumed." What does this cartoon convey about the nature of scientific discovery? *Source:* Courtesy of Abraham Tulp.

that they are introducing new ideas and struggling with what is not already obvious. Rarely is one side completely "wrong," although they are typically portrayed as such by their opponents! Disagreement, when coupled with constructive dialogue, means that scientists are working towards a fuller, more objective account than either has yet provided. Far from being a weakness, controversy often reflects a healthy, growing science.

QUESTIONS AND ACTIVITIES

1. What does this case show about the following aspects of doing biology?
   — the role of dead ends, blind alleys, and revisions to theory
   — cumulative versus revolutionary growth of knowledge
   — the value of novel findings
   — the collective versus individual nature of scientific inquiry and discovery
   — disagreement
2. Identify several assumptions about the nature of science conveyed in the cartoon image in Figure 8.4. Name at least one other account of a scientific discovery where one or more of these assumptions is evident. Based on this chapter and on other cases in this book, how might you revise the cartoon portrayal? Describe at least two other alternative images or metaphors for science.

3. Scientists generally receive recognition from their peers for novel findings. Many consider this a reward structure that motivates scientists to make important discoveries. Some feel, however, that this creates an unhealthy competitive atmosphere and promotes hasty rather than careful work. Describe how else you might promote creative discovery and original work in science. How do you think scientists should be given credit for their work?

4. Many lethal pesticides, such as dinitrophenol (DNP) and dichlorodiphenyltrichloroethane (DDT—see Chapter 17), can carry protons across a membrane that otherwise does not allow protons to pass through. Explain how these pesticides will affect energy processing in the cell.

5. Many chemists agreed that Jagendorf's acid-bath experiments showed how an imbalance of protons *could* produce ATP. But they did not agree that it was necessary. How might they have argued this?

SUGGESTED READING

Gilbert, N., and M. Mulkay. 1984. *Opening Pandora's Box*. Cambridge, England: Cambridge University Press.

Harold, F. 1986. *The Vital Force*. New York: W. H. Freeman & Company.

Robinson, J. 1984. "The Chemiosmotic Hypothesis of Energy Coupling and the Path of Scientific Opportunity." *Perspectives in Biology and Medicine* 27: 367–383.

Robinson, J. 1986. "Appreciating Key Experiments." *British Journal of the History of Science* 19: 51–56.

Weber, B. 1991. "Glynn and the Conceptual Development of the Chemiosmotic Theory: A Retrospective and Prospective View." *Bioscience Reports* 11(6): 577–617.

CHAPTER 9

Walter Cannon & Self-Regulation in Animals

JOEL B. HAGEN

☐ INTRODUCTION

In 1917, the United States entered World War I. Joining in a European conflict was a controversial decision for a traditionally isolationist nation, but the "Great War" also stimulated a wave of patriotism among Americans. One volunteer was Walter Bradford Cannon, a medical researcher at Harvard University. He could have easily avoided military service because of his age and his family responsibilities—at 46, he had a wife and five dependent children. Nevertheless, Cannon enlisted in the army and joined a medical unit stationed near the front lines in Belgium. Here he encountered the horrors of the modern battlefield. It was, as he later described, "a scene so awful that it seemed to me almost beyond belief that in the midst of it were *men*, with eyes and ears and sensitive nerves, who were being ripped open and mangled as they endured a maelstrom of tumult and horror."

Aside from his clinical duties, Cannon studied the causes of physiological shock. Soldiers who initially survived their wounds often died from this disorder, which is characterized by a rapid drop in blood pressure. This was particularly likely to occur in war, where casualties bled profusely. Working in the shock ward of a field hospital was a wrenching experience. Cannon vividly described the ghastly wounds, the filth, the delirious cries of injured men, and the frequent deaths. In the years before blood or plasma transfusion had been perfected, Cannon and other physicians could do little to reverse the deadly effects of shock.

These gruesome wartime experiences might impress a scientist with the frailty of life—but also its resilience. How do animals survive in an often dangerous world? Why doesn't every injury lead to a fatal disruption of vital processes? How is the normal balance of life maintained in a constantly changing environment? Such questions had interested Cannon even before World War I, and they remained the primary focus of his scientific career.

THE IDEA OF SELF-REGULATION

On the walls of Cannon's office hung portraits of two great scientists: Charles Darwin and Claude Bernard. Like most biologists, Cannon was heavily influenced by Darwin's evolutionary theory, but he owed an even greater intellectual debt to Bernard. In a famous set of lectures delivered at about the time Cannon was born, the famous French physiologist described the difference between an organism's *internal environment* and its *external environment*. The internal environment was the blood and other fluids surrounding the body's cells. Although the external environment is constantly changing, the internal environment remains remarkably stable. Bernard concluded that the internal environment served as a kind of buffer between living cells and the fluctuating external environment.

These ideas proved to be extremely influential, and they served as guiding principles for Cannon's research. Late in his career, Cannon revised Bernard's suggestive ideas into a much more detailed form. In his popular book, *The Wisdom of the Body*, Cannon coined the term **homeostasis** for the modern concept of biological self-regulation. He based his concept on a large body of experimental evidence that he and other physiologists had gathered. In contrast to Bernard, who was unimpressed by Darwin's theory, Cannon described homeostasis as an important evolutionary adaptation. Cannon's idea became a central concept in physiology, and it was later borrowed by scientists in several other fields as well.

THE MAKING OF A SUCCESSFUL PHYSIOLOGIST

Like many successful scientists, Cannon began doing research in college. During his senior year at Harvard, he wrote his first scientific article reporting the results of experiments on how microscopic organisms orient toward light. The next year, as a first-year medical student, Cannon began experimenting with X rays, which had been just discovered. He found that X-ray images of the digestive tract became much clearer after the patient drank a solution of barium. This innovation was quickly adopted by physicians, and it remains a standard procedure today.

Although Cannon planned to become a physician when he entered medical school, he found his clinical courses dull. His early successes in the laboratory encouraged him to pursue a career in research. Some of this research, including his wartime studies of shock, was aimed at treating seriously injured patients. But Cannon was more interested in understanding the physiology of healthy organisms. These experimental studies provided the foundation for his concept of homeostasis.

NERVES, HORMONES, AND SELF-REGULATION

Cannon's early experiments with X rays led to an interest in digestion and how it is controlled. As a medical student, he spent hours in front of the X-ray screen observing rhythmic contractions (peristalsis) in the digestive tracts of dogs, cats, and other animals. He noticed that whenever an animal became excited, the contractions stopped. Was this response controlled by the nerves or by a hormone?

Both hypotheses were reasonable, but peristalsis turned out to be more complex than it appeared. In an early experiment, Cannon drew blood from a cat both before and after it was exposed to a barking dog. Before the stressful encounter, Cannon detected no adrenal hormones in the cat's blood. Almost immediately after exposure, however, an adrenal hormone—later called epinephrine—could be detected. Cannon then tested the two blood samples on a small strip of intestinal muscle. Isolated from the body and suspended in a dilute salt solution, the muscle continued to contract rhythmically. When blood containing epinephrine was applied to the muscle, peristalsis ceased, but when the muscle was rinsed and exposed to epinephrine-free blood, the rhythmic contractions began again.

Could this response to epinephrine be demonstrated in living animals? Was the secretion of epinephrine controlled by the nervous system? Cannon tried to answer these questions by cutting the sympathetic nerve leading to one adrenal gland but leaving intact the nerve to the other gland. After the intact nerve was electrically stimulated or after the cat was exposed to a barking dog, Cannon detected epinephrine in the cat's blood. Later the experimental cat was killed and the weights of the two adrenal glands compared. Whenever this experiment was done, the gland without neural connections always weighed more than the gland with intact nerves. Cannon concluded that the loss of weight was due to the epinephrine secreted by the intact adrenal gland.

PROBLEM

What alternative conclusions could be drawn from Cannon's results? Explain whether Cannon's surgical procedure was a controlled experiment. How could the experiment have been designed differently?

An active man who enjoyed competitive sports, Cannon turned to an athletic metaphor when he described his experimental results. Referring to the combined physiological regulation as a form of teamwork, he emphasized that the nervous and endocrine systems work together. Just how extensive was this teamwork?

Perfecting his surgical technique, Cannon and his students later removed sympathetic nerves leading to several other important organs, including the heart and the liver. Deprived of its sympathetic nerves, the heart continued to beat rhythmically. Even without neural stimulation, however, the heart responded to hormones. For example, when the adrenal glands were artificially stimulated, the heart rate rapidly increased. Conversely, injecting large amounts of insulin caused a rapid decrease in heart rate and a drop in the level of blood sugar. These effects were quickly reversed by epinephrine (and, as we now know, glucagon), which mobilized sugar from the liver and increased heart rate.

Cannon found that the liver continued to perform its function as a reservoir of sugar even when all nerves to the organ were severed. If the nerves to the adrenal glands were also cut, however, cats almost always went into fatal convulsions after being injected with insulin. Similar results occurred after chilling animals, either by placing them in cold environments or by injecting ice water into their stomachs. As long as the adrenal glands were connected to the sympathetic nervous system, cats responded adaptively to the cold stress. They shivered, fluffed their fur to increase insulation, and became more active. When nerves to both the liver and the adrenal glands were cut, the cats usually died.

The final step in this line of experimentation involved destroying the entire sympathetic nervous system. Cannon removed all of the ganglia and their interconnecting nerves (Figure 9.1). The technique had been tried before, but the experimental animals almost always died after surgery. This led many physiologists to conclude that the sympathetic nervous system was essential for survival. Cannon came to a slightly different conclusion after he discovered that his "sympathectomized" animals

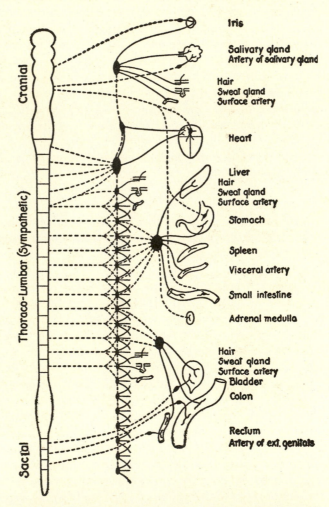

FIGURE 9.1 A diagrammatic representation of the sympathetic nervous system. Clusters of nerve cell bodies (ganglia) lie near the spinal column. The axons of these nerves lead to nearly all of the organs of the body. By removing the ganglia, Cannon was able to destroy the action of the entire sympathetic nervous system. *Source:* from *The Wisdom of the Body*, Revised Edition by Walter B. Cannon, M.D. Copyright 1932, 1939 by Walter B. Cannon, renewed © 1960, 1967, 1968 by Cornelia J. Cannon. Reprinted by permission of W. W. Norton & Company, Inc.

often led remarkably normal lives. Several of Cannon's cats continued to live for months or even years, and at least one female gave birth to healthy kittens. If the animals were removed from the protected environment of the laboratory, however, their condition often changed dramatically. Unlike normal cats, the sympathectomized animals could not respond adaptively to stress. When forced to do even moderate exercise, the cat's heart rate often failed to increase in response to the body's demand for more oxygen. When emotionally excited—for example, by being placed near a barking dog—the cat's blood sugar level failed to increase as it would in normal individuals. When sympathectomized cats breathed an atmosphere with reduced oxygen (as might occur at a high altitude), they usually fainted. It seemed that all of the "fight or flight" responses that allow animals to respond adaptively to emergencies were under the primary control of the sympathetic nervous system. Without this protective regulatory system, experimental animals could survive only if they were artificially protected from all types of physiological stress (see Chapter 10).

THE CONCEPT OF HOMEOSTASIS

Cannon completed this series of surgical experiments just before he wrote *The Wisdom of the Body* (1932). Although he had used the term *homeostasis* a few times before, Cannon's book made it famous. Written for a broad scientific audience, the book generated so much interest that it was revised and republished in 1939. Thirty years later, it was reprinted again in paperback form.

The title of Cannon's book may appear odd. Are subconscious neural impulses and the secretion of hormones really a form of "wisdom"? Cannon described how homeostasis had evolved through a process of trial and error. Over the course of millions of years, many species went extinct, but through natural selection successful species had evolved ways to regulate biological processes. This evolutionary process culminated in the exquisite system of hormones and autonomic nerves that allows mammals to maintain a high degree of internal stability. Virtually every important physiological function could be regulated, including body temperature, pH, amounts of water and salts, the level of sugar in the blood, and the metabolism of sugar and fat within the cells of the body. Through the course of evolution, the body had "learned" to regulate itself.

Cannon emphasized the remarkable precision of homeostasis. Underlying this optimistic theme, however, was a more sober recognition that self-regulation is not a perfect adaptation. His work as a physician, particularly his battlefield experiences during World War I, reminded Cannon that severe injuries can often overwhelm the body's self-regulatory mechanisms. His experiments on cats had also vividly demonstrated that if the endocrine or nervous systems are damaged, homeostasis is easily disrupted. How could Cannon integrate these two contrasting themes in his book?

Through the process of natural selection, Cannon believed that well-adapted species had evolved very flexible control systems. Like an engineer who designs a bridge to withstand forces greater than those normally encountered, natural selection had produced wide margins of safety in the body. Organisms could usually tolerate large fluctuations in almost any physiological function, at least temporarily. Various sense receptors acted as early warning devices to alert the body to dangerous changes. Then, through the teamwork of nerves and hormones, the body counteracted these changes. Homeostasis was never perfect, but under most conditions, the body maintained a high degree of internal constancy.

ANIMALS, MACHINES, AND SOCIETY

Together with evolution and cell theory, homeostasis became one of the central organizing principles in modern biology. Building on the foundation of Cannon's *Wisdom of the Body*, physiologists used homeostasis to help explain the function of every organ system of the body. Homeostasis has been used by other biologists to describe self-regulation in a wide variety of living systems, including developing embryos, cancerous tumors, populations, and even ecosystems. In the years after the publication of his most famous book, Cannon's influence also stretched far beyond biology.

After World War II, the mathematician Norbert Wiener recalled how Cannon influenced his thinking about self-regulation. Wiener joined a number of young scientists who attended a weekly dinner seminar led by the famous physiologist. The seminar was a place where new ideas were presented, discussed, and criticized by the members. Wiener later helped to establish the new field of cybernetics, which studied the general principles of self-regulation. He developed the mathematical idea of **negative feedback** and applied it to machines as diverse as anti-aircraft guns, radios, and thermostats. Many biology textbooks today use negative feedback and mechanical models (for example, the thermostat) to explain homeostasis. For Wiener, who credited Cannon with inspiring some of his ideas, there was really no difference between self-regulation in animals and machines. Was such a generalization warranted? Should scientists try to extend theories from one field to another?

Cannon recognized that there is a danger in applying theories too broadly. "Nothing is easier than to let one's imagination spin fancies on the basis of slight evidence," he warned younger biologists. Yet Cannon himself was sometimes willing to go on a limb. In the final chapter of *The Wisdom of the Body*, he left his specialty of physiology to discuss what he called "social homeostasis."

As he was writing the chapter, the nation was suffering the worst economic catastrophe in its history: the Great Depression. In some industrial cities, more than half of all workers were unemployed. Prices for agricultural products dropped so precipitously that many farmers could no longer make mortgage payments. Angry mobs attacked judges and bankers who ordered foreclosures and evictions. Radicals on both the right and left predicted the collapse of American democracy. Was the body politic suffering something analogous to the traumatic shock that Cannon had witnessed on the battlefields of World War I?

Cannon drew analogies between the economic plight of the nation and the physiological crises faced by living organisms. Like the highly adapted mammalian body, Cannon believed that industrial societies were homeostatic. Under normal circumstances, the political and economic systems were capable of regulating themselves to achieve a stable balance. Like biological homeostasis, however, social regulation came at a high cost. In order to maintain stability, societies needed to allocate economic resources to government programs that promoted harmony and social stability. Although he was a lifelong Republican, Cannon supported many of Franklin Roosevelt's New Deal programs, believing this government intervention was part of the homeostatic mechanism that would bring an end to the Great Depression and restore social harmony.

Cannon's discussion of social homeostasis is important because it serves as a reminder that laboratory life is seldom completely isolated from the broader social lives of scientists. Cannon was a cautious experimentalist who knew about the dangers of scientific speculation. Yet he was also an imaginative and creative thinker who was willing to take intellectual risks. His strong sense of social responsibility prompted him, after some hesitation, to speak out on the political and economic issues of his day. Careful experimentation and intellectual risk taking: Cannon's legacy as a great biological thinker rested on his ability to balance these two seemingly contradictory characteristics.

☐ EPILOGUE

Throughout his career, Cannon was forced to defend his use of cats and other animals as experimental subjects. To an even greater degree than the Great Depression, the issue of animal experimentation thrust him into the political arena. Although there is no evidence that Cannon was needlessly cruel to his laboratory animals, his experiments often resulted in their deaths. For many people, such animal experimentation was unacceptable. This "antivivisection" controversy came to a head when his friend and former teacher, William James, wrote a letter criticizing Cannon's research. Antivivisectionists publicized the letter, and it was widely read.

Although James is best known today as a philosopher and psychologist, he was trained as a biologist, and early in his career he taught anatomy and physiology. When he wrote the antivivisection letter at the end of his long career, James was one of the most respected intellectual figures in the United States. His criticism of Cannon, therefore, carried both conviction and authority.

Antivivisectionists were critical of animal experimentation for several reasons. Pain and suffering of animals was certainly an issue, but not the only one. James knew that Cannon was a careful surgeon, that he used anesthetics, and that his experimental animals were cared for properly. Of greater concern was the potential dehumanizing effects that animal experimentation might have on scientists. Routine killing, even when done humanely, might cause experimentalists to become less sensitive to pain and suffering. It might also lead some scientists to callously accept human experimentation. After all, if it was all right to experiment on nonhumans, couldn't one also justify vivisection on prisoners, the mentally retarded, or members of minority groups if the results might benefit the general public? Finally, James criticized scientists for being too authoritarian and resisting any attempts by nonscientists to oversee their work. If experimental science were really being done for the public good, shouldn't the public be allowed to regulate the types of experiments done by scientists?

Cannon considered such criticisms to be irrational, as do many biologists today. The issues have not disappeared, however. Indeed, they have become even more sharply defined. This is partly the result of the rapid growth and high visibility of science after World War II. Experimental animals are now used in research, product testing, and education on a very large scale. Reliable estimates place the number of vertebrate animals used for these purposes in the United States at 25 to 30 million per year. Public attitudes have also changed as a result of sensational cases of inhumane treatment of animals by just a few scientists. For example, in a notorious case at the University of Pennsylvania, researchers inflicted traumatic head injuries to unanesthetized monkeys and baboons. Videotapes of the experiments provided graphic evidence that some of the scientists seemed to enjoy inflicting pain on the helpless animals. Although this research was probably aberrant, it again raises the issue of whether animal experimentation may cause scientists to become callous toward their subjects. More recent revelations of radiation experiments done on mentally retarded children after World War II by doctors supported by the Atomic Energy Commission revived the antivivisectionists' long-standing fear that the widespread acceptance of animal experimentation may lead people to accept human experimentation.

The animal rights movement has seized on such cases to publicize and promote its goals. Although much of the literature of this popular movement is marked by overstatement and emotional appeals, serious ethical issues concerning the use of animals in science continue to be raised. Professional philosophers (Peter Singer and Tom Regan) and even some scientists (Richard Ryder and Mary Dawkins) have written thoughtful books and articles critical of animal experimentation. Singer argues that when the benefits of science are balanced with the suffering of experimental subjects, animals and humans must be treated equally. A benefit to humans, no matter how great, cannot justify the suffering and death of just animals alone. Only if both the suffering and the benefits are equally shared among all of the species involved can vivisection be justified. Regan makes the even stronger case that all sentient animals share certain rights, the most important of which is the right to life. These arguments have been criticized by other philosophers (Carl Cohen, Michael Leahy, and Raymond Frey), who claim that equating human and nonhuman interests is unwarranted.

These disagreements show no signs of being resolved, but they have focused attention on serious practical issues concerning the social responsibilities of scientists. Between the extreme positions in the animal rights controversy is a growing recognition that society has a legitimate role in guiding scientific research, particularly when it is supported by public funds. Even if some animal experimentation is necessary, scientists should be aware of important ethical issues raised by their work.

QUESTIONS AND ACTIVITIES

1. What does this case show about the following aspects of doing biology?
   — relationship between experiments and theory
   — boundaries of scientific disciplines
   — social, ethical, and political responsibilities of scientists

2. Cannon was primarily interested in how homeostasis maintains normal physiological balance. Describe how homeostasis can also explain pathological conditions such as the traumatic shock that Cannon encountered as a physician during World War I.

3. Discuss how your biology textbook uses negative feedback to explain homeostasis. Is self-regulation in machines and organisms basically the same, as Norbert Wiener claimed, or are there some important differences?

4. Discuss whether Cannon was justified in drawing analogies between physiological homeostasis and "social homeostasis." What are the possible benefits and dangers in this type of analogical reasoning?

5. Examine your school's guidelines for animal experimentation. To what extent do the guidelines balance public accountability with the legitimate goals of scientific research? Discuss whether Cannon's experiments would be allowed at your school. If not, could they be redesigned to make them acceptable?

SUGGESTED READING

Allen, G. E. 1975. *Life Science in the Twentieth Century*. New York: John Wiley & Sons.

Bennison, S., A. C. Barger, and E. L. Wolfe. 1987. *Walter B. Cannon: The Life and Times of a Young Scientist*. Cambridge, MA: Harvard University Press.

Bennison, S., A. C. Barger, and E. L. Wolfe. 1991. "Walter B. Cannon and the Mystery of Shock: A Study of Anglo-American Co-operation." *Medical Research* 35: 217–249.

Cannon, W. B. 1939. *The Wisdom of the Body* (revised edition). New York: W. W. Norton.

Cannon, W. B. 1945. *The Way of an Investigator: A Scientist's Experiences in Medical Research*. New York: W. W. Norton.

Cross, S. J., and W. R. Albury. 1987. "Walter B. Cannon, L. J. Henderson, and the Organic Analogy." *Osiris* 3: 165–192.

Fleming, D. 1984. "Walter B. Cannon and Homeostasis." *Social Research* 51: 609–640.

Orlans, F. B. 1993. *In the Name of Science: Issues in Responsible Animal Experimentation*. Oxford, England: Oxford University Press.

CHAPTER 10

Hans Selye, Hormones, & Stress

FRED SINGER

☐ *INTRODUCTION*

Many people who enjoy a good chicken dinner are not aware that the plumpest, most tender meat comes from roosters that were castrated when they were young. Someone probably discovered this accidentally, when they castrated a rooster to reduce its overall activity. Arnold Berthold wondered how removal of the testicles brought about these anatomical and behavioral changes. Did the testicles somehow interact with the nervous system to bring about normal development and behavior? Alternatively, maybe they produced some substance that was essential for normal development and behavior. If so, what was this substance, and how did it exert its control?

In 1848, he castrated two young roosters and transplanted their testicles into their abdominal cavities. Rather than maturing into fat, tender, and inactive capons, the birds behaved normally, defending their territories, crowing and strutting with all the vigor and enthusiasm of true roosters. Berthold autopsied these animals and found no evidence of nerve regeneration but discovered a rich network of capillaries connecting the testicles and the circulatory system. He concluded that the testicles were contributing something to the blood via this capillary network that affected rooster behavior and anatomy.

In subsequent decades, organ removal studies similar to those conducted by Berthold revealed a great diversity of chemical messengers—hormones—that are carried in the blood. Hormones are released by many glands and organs and may stimulate or inhibit the activities of sensitive tissues or organs. By the early twentieth century, a great hormones search was in full swing. In a modification of Berthold's technique, physiologists removed glands suspected of being sources of hormones, ground them up, and added solvents to extract the suspected hormones from the glands. These extracts were then injected into test animals, which were studied for anatomical, physiological, or behavioral changes. This blind trial-and-error technique sometimes led to discovery of new and important hormones and, other times, led to dead ends.

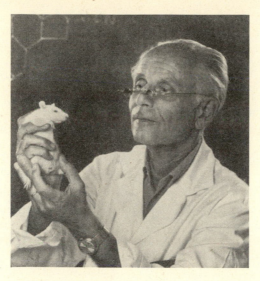

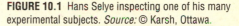

FIGURE 10.1 Hans Selye inspecting one of his many experimental subjects. *Source:* © Karsh, Ottawa.

In at least one case, an investigator did not accomplish his intended mission. Instead, Hans Selye opened up an entirely new field of inquiry only partly related to his initial experiment.

HANS SELYE

Hans Selye (Figure 10.1) entered endocrinology when the field was in its heyday. Following a long family tradition, Selye became a physician. Selye was impressed by his father's technical expertise and moved by his sense of humanity and compassion. If the patient was poor and unable to pay for a physician's services, rather than charge for the visit, Selye's father would often leave a small sum of money behind. Years later, Selye decided to forsake a clinical career in favor of a life devoted to medical research. This decision was not made lightly. Selye wrote with considerable sadness that he would never have the satisfaction of seeing the grateful eyes of a mother whose child has been saved from certain death.

Selye grew up in a multicultural family. His father was Hungarian, his mother Austrian, and his two governesses English and French. After the shifting of national boundaries following World War I, his hometown became Czechoslovakian. Thus he could boast of having six native tongues, to which he added several others during the course of his lifetime.

As a result of his dynamic personality and his ability to communicate in many languages, Selye was a very charismatic figure. He was a brilliant lecturer who could draw two different pictures on the blackboard using both hands simultaneously while carrying on a conversation about what both hands were doing. His powers of persuasion were unparalleled, in part because he was fluent in so many languages and in part because he was a prolific writer, authoring 38 books and approximately 1,700 articles over the course of his lifetime. He dazzled visitors by inviting them into his state-of-the-art laboratory, performing surgery on rats while the visitors watched, and discussing the significance of his findings in whatever language they spoke best.

Selye first encountered the problem that dominated his career as a second-year medical student at the University of Prague. His professor presented the medical students with five patients suffering from different diseases. The professor carefully questioned and examined each patient as the future physicians looked on. Though each patient suffered from a different disease, they shared several common symptoms, such as a coated tongue, aches and pains in the joints, intestinal disturbances, and appetite loss. Most patients had a fever; some had skin rashes and inflamed tonsils, spleens, and livers. Selye's professor ignored these shared symptoms, instead focusing on how certain symptoms were diagnosed; for example, little red-and-white "Koplik spots" on the inside of the cheeks near the molars indicated measles.

Selye, in contrast, was very excited about the symptoms that characterized "the syndrome of just being sick," as he originally called it. What caused this common suite of physiological responses? He wondered why nobody else seemed to pay any attention to this syndrome. Of course, there was no way a 19-year-old medical student could convince anyone to pay attention to his finding, which most people regarded as trivial and unworthy of serious consideration.

SEARCHING FOR NEW OVARIAN HORMONES

Selye's insight became more important a decade later in 1936, as he began his career as a postdoctoral biomedical researcher at McGill University in Montreal. His professor, J. B. Collip, was searching for new ovarian hormones, so Selye's first task was to retrieve a bucket of fresh cow ovaries from the slaughterhouse. Collip ground the ovaries and prepared extracts with different solvents. Selye then tested these extracts on some female rats.

A few days after injecting his first rats with one of the extract solutions, Selye killed the rats and performed autopsies, hoping to find anatomical changes. If the treated rats showed changes in response to the extract treatment, he would have evidence for a new hormone. Much to his joy, he discovered the following triad of anatomical changes:

1. considerable enlargement of the adrenal cortex (outer layer of the adrenal gland, located just above the kidneys),
2. intense shrinking of the lymphatic structures, including the thymus, spleen, and lymph nodes, and a major reduction in the number of eosinophils (a type of white blood cell associated with immune response),
3. deep ulcers in the lining of the stomach and duodenum (Figure 10.2).

No ovarian hormone had ever been shown to induce these changes, and Selye was enthusiastic about his good fortune. Next, he tested extracts of the placenta which had already been shown to produce other sex hormones. He discovered the same anatomical changes. When he tried extract of the pituitary gland (located at the base of the brain), he was astounded to find the same effect. Why should three different glands produce the same hormone?

One hypothesis was that the rats were reacting not to a specific hormone in the extract, but rather to the pain and stress of being handled, jabbed with needles, and kept under less than ideal conditions by an inexperienced researcher. As an endocri-

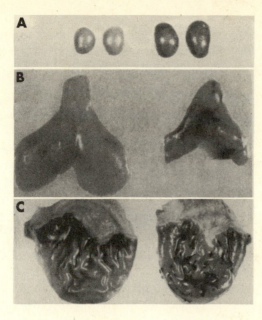

FIGURE 10.2 Common anatomical response to stress during the alarm reaction. The pictures on the left are from a normal rat, while those on the right are from a rat that was immobilized on a board for 24 hours prior to being sacrificed. Notice the (A) enlargement of the adrenals, (B) shrinking (involution) of the thymus, and (C) pitting and ulceration of the gastric mucosa. *Source:* from Hans Selye, *The Story of the Adaptation Syndrome.*

nologist embarked on the great search for a new ovarian hormone, Selye was not enthusiastic about this hypothesis. Yet it needed to be investigated.

How could he test this hypothesis? If his rats were showing a response to pain and stress, then any unpleasant stimulus should produce the same effects. He subjected six rats in each experimental group to a different potentially stressful stimulus, leaving one group of six rats as a control. The accompanying table shows the response of thymus and adrenal glands to stressful stimuli.

Treatment	Mean Thymus Weight (mg)	Mean Adrenal Weight (mg)
Control (untreated)	281	50
Starved for 96 hours	80	56
Exposed to 5°–7° C for 48 hours	102	65
Legs tied for 48 hours	82	68
Extensive skin lesions	104	45
Fracturing both tibias and femurs	145	50
Removing intestines from body for 1 minute	100	66
Atropine (2 cc of 1% solution, 4 times over 2 days)	55	64
Morphine (2 cc of 1% solution, 4 times over 2 days)	78	49
Epinephrine (0.1 cc of 0.1% sol., 4 times over 2 days)	84	44
Formaldehyde (0.5 cc of 4% sol., 4 times over 2 days)	64	60

PROBLEM

In what ways do these results support or fail to support Selye's hypothesis? A second prediction of Selye's hypothesis might be that more unpleasant treatments will cause more pronounced effects. Do the data support or refute this prediction?

Selye was seriously depressed. He had been looking for an ovarian hormone, but almost any treatment caused the same general effect. Then he flashed back to the human patients he had observed in medical school, all of whom shared the "syndrome of just being sick." Perhaps these rats were showing a similar syndrome—some sort of physiological response to pain or stress. Maybe his rats and the sick humans were showing the same types of physiological reactions to general trauma.

Selye was particularly excited by the idea that this response of the body to all types of stress might actually be adaptive or useful—in essence, be Nature's way of fighting disease and injury. Perhaps, with future research, physicians could help this defensive process along by promoting the response, and thus reduce the suffering that people go through when they are sick.

Most of Selye's associates were unimpressed with this line of research. In fact, Collip called Selye into his office for a heart-to-heart chat on the problems of this research plan. Collip explained that Selye had the potential to be a successful endocrinologist, so why was he wasting his time looking for a general effect that could be caused by any substance or treatment? As Collip stated emphatically, "But, Selye, try to realize what you are doing before it is too late. You have now decided to spend your entire life studying the pharmacology of dirt!"

Despite a lack of enthusiasm by his colleagues, Selye persisted and, in fact, devoted the rest of his life to studying this process, which he called the **stress response**. One of the most difficult, challenging, and stimulating problems faced by a scientist is to decide what to do with a new idea. Is the idea really worth pursuing? If so, how should it be pursued? What are the important questions? What avenues of investigation will be fruitful, and what avenues will lead to dead ends?

Selye knew that some of the affected structures had already been shown to be important sources of hormones. He reasoned that if these anatomical changes affect hormone release, then there should be behavioral or physiological responses to stress. He also knew from his autopsy studies that the anatomical changes often went away, even if the source of stress continued. In fact, the animals subjected to continued stress appeared to be more resistant to additional stress. He wanted to know whether his reasoning and impressions were correct. Specifically, how does resistance to stress change over time? Also, does exposure to a stressor make the individual more resistant to other sources of stress?

In an experiment designed to answer these two questions, Selye used moderate cold (2° C) as the first stressor, and extreme cold (-4° C) as a second source of stress. He autopsied 10 rats from each group; all cold-treated rats showed the triad of anatomical changes, while none of the control group showed any anatomical changes. Selye then transferred 20 of the remaining 90 cold-treated rats to an even colder chamber (-4° C) along with 20 of the remaining controls. The accompanying table shows the experimental design.

Time	N	Experimental Group	N	Control Group	Experimental Comparison
0	100	Keep at 2° C for 48 hours	100	Keep at 20° C for 48 hours	
48 hours	10 20 70	Autopsy Move to -4° C Stay at 2° C for 5 weeks	10 20 70	Autopsy Move to -4° C Stay at 20° C for 5 weeks	Compare anatomy Compare survival
5 weeks	20	Move to -4° C	20	Move to -4° C	Compare survival
Until Death	50	Stay at 2° C	50	Stay at 20° C	Compare survival

N = number of rats receiving each treatment

Selye found that the control rats survived much better than the cold-treated rats following the transfer to -4° C.

PROBLEM

Propose a hypothesis to explain why the control rats survived better when moved into the subzero temperatures at 48 hours.

Selye wanted to know whether the same pattern would hold after a longer continuous exposure to the 2° cold treatment. His early studies had shown that the triad of anatomical changes was in some cases reversed by five weeks of constant exposure to stress. Would rats exposed to constant cold for a long period of time be more resistant than rats exposed to cold for 48 hours? After exposing his experimental rats to 2° C for five weeks, Selye transferred 20 more of these cold-treated rats and 20 untreated controls to -4° C. In contrast to his findings at 48 hours, Selye found that the cold-treated rats had a much higher survival rate than the controls following the transfer to -4° C. What could account for these findings?

Was there some long-term cost to the rats from being continuously exposed to the stressful stimulus? Selye allowed his 50 surviving cold-treated rats to live out their remaining lives at 2° C, while his 50 remaining untreated controls continued a more normal 20° C existence. He found that the cold-treated rats had much shorter life spans than the untreated controls.

THE GENERAL ADAPTATION SYNDROME

On the basis of this and other experiments conducted during the mid- and late 1930s, Selye constructed a three-stage model of physiological response to pain or stress that he called the **General Adaptation Syndrome (GAS)**. The first stage is the *alarm reaction*, in which the body makes its initial response to the stressful stimulus. Anatomical changes during this period include the triad of responses Selye discovered in his first experiments (Figure 10.2). Other researchers stimulated by his findings replicated his experiment and were able to show a more consistent increase in the adrenal gland than Selye had shown. Selye also found major physiological changes in the early part of the alarm reaction (which he called "the shock phase"),

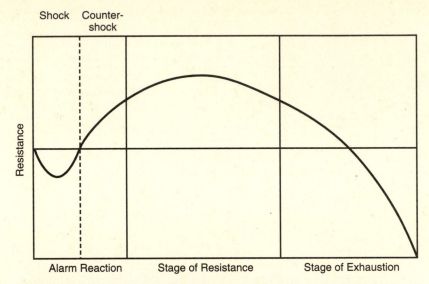

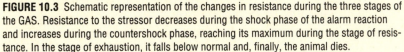

FIGURE 10.3 Schematic representation of the changes in resistance during the three stages of the GAS. Resistance to the stressor decreases during the shock phase of the alarm reaction and increases during the countershock phase, reaching its maximum during the stage of resistance. In the stage of exhaustion, it falls below normal and, finally, the animal dies.

including reduction in blood pressure, hypothermia, decrease in muscle tone, and depression of the nervous system. Most of these anatomical and physiological changes are reversed during the latter part of the alarm reaction ("countershock phase"), after the body begins to respond to the hormones that are released into the bloodstream (see Figure 10.3 for a graphical representation of the GAS).

If the source of stress persists, the body enters the second stage of the GAS, which is the *stage of resistance* or *stage of adaptation*. During this stage, the body shows increased resistance to the particular stressor to which it is being exposed. The triad of anatomical responses disappears or is greatly reduced; thus, the animal has adapted to the stressful stimulus.

With continued exposure to the same stressor, the animal enters the final stage of the GAS, which is the *stage of exhaustion*. During this stage, the animal experiences a reoccurrence of the symptoms of the alarm reaction and appears to lose the adaptation it has developed in the second stage. If the stage of exhaustion continues for too long, the animal dies.

PROBLEM

Use Selye's GAS model to interpret the findings of his experiments with cold-stressed rats. In which stages of the GAS were the cold-stressed rats during each part of Selye's experiment?

The GAS describes what happens during stress without really explaining why these changes occur. Selye proposed that hormonal changes were responsible for the results of his experiments with stress. Based on the observed anatomical changes, he hypothesized that the adrenal and thymus glands released hormones that controlled the stress response.

Why did animals lose their resistance during the final stages of the GAS? Selye proposed that, at birth, individuals have a finite quantity of "adaptation energy," which is gradually consumed by exposure to a stressor. When all the adaptation energy is used up, the animal dies. Selye proposed that adaptation energy can be partially restored by rest and healing but that an individual's life span is shortened by periods of stress.

PROBLEM
How could Selye test his concept of adaptation energy? What challenges are associated with designing such a test?

How Does the Stress Reaction Work?

Selye provided a description of several different changes that occur in the body in response to different types of stressors. His challenge was to generate an accurate picture of what actually happens within the body of a stressed animal.

When Selye was doing his work, the adrenal medulla had already been shown to secrete the hormone epinephrine (also known as adrenaline), which raised blood pressure in response to stressors (see Chapter 9). He knew that the medulla had some role in the very early part of the shock phase. But the increase in the size of the adrenal gland during the alarm stage was due to an increase in the size of the cortex of the adrenal gland—not the medulla. Selye wanted to know if the increase in the adrenal cortex caused the decrease in the thymus.

This idea was stimulated, in part, by work in several other laboratories, where researchers were feverishly attempting to characterize the hormone produced by the adrenal cortex. This unidentified hormone—initially named "cortin"—was produced by grinding up a cow's adrenal cortex and putting the grounds through a series of solvents. This extract was then injected into research animals. As research on this substance continued, many techniques of purification were developed, but the active ingredient itself was still unknown. Using the available technology, Selye conducted a series of experiments to determine more precisely the adrenal gland's role in the stress response.

Selye removed the entire adrenal glands of a group of rats and left the adrenal glands intact in several groups. He subjected the rats to stressors of various types; in addition, all rats except the untreated controls were starved during the course of the experiment. Selye discovered that rats without adrenal glands showed only a slight reduction in thymus size. Two groups of rats were given high doses of cortin or epinephrine (adrenaline in Figure 10.4). Even these rats showed only a slight reduction in thymus size.

PROBLEM
How do the findings in Figure 10.4 support or refute Selye's claim that hormones produced by the adrenal glands shrink the thymus gland during the stress response? Design your own experiment to investigate this question.

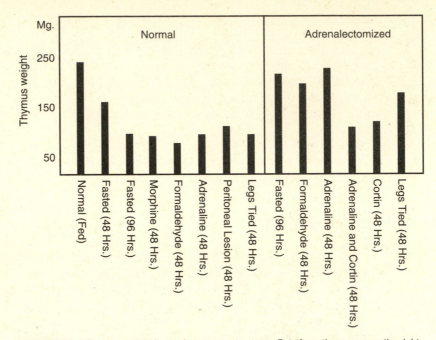

FIGURE 10.4 Thymus weight in rats in response to stress. Rats from the groups on the right had their adrenal glands removed before being subjected to food deprivation. Two of these groups of rats were injected with adrenal gland hormonal extracts. *Source:* Hans Selye, "The General Adaptation Syndrome and the Diseases of Adaptation," *Journal of Clinical Endocrinology and Metabolism* 6: 117–230.

Diseases of Adaptation

As Selye describes in his writings, part of the motivation for his career choice was a humanistic interest in eliminating disease. He hoped that understanding the stress response would allow him to help eliminate or reduce the trauma of certain diseases.

Many of the symptoms of the stress reaction were very similar to symptoms of diseases that were commonly afflicting people during the first part of the twentieth century. For example, animals during the stage of exhaustion sometimes developed serious kidney diseases, often accompanied by inflamed arteries, proteins excreted in the urine, high blood pressure, and the formation of nodules on the heart tissue. Selye hypothesized that these diseases were actually diseases of adaptation caused by excessive production of cortical hormone in response to stress. While high cortical production allowed the animal to deal with short-term stress, over the long haul, continued cortical production exhausted the animal's adaptation energy, leading to serious disease and possibly death.

Selye reasoned that if he was correct he should be able to induce symptoms of diseases of adaptation by injecting rats with high levels of cortical extract. Biochemists had recently developed techniques for producing significant amounts of deoxycorticosterone acetate (DCA), a hormone that affects salt and water metabolism. Selye administered high doses of DCA to experimental rats, fed them high-salt diets, and removed one of their kidneys. In response to this treatment, rats developed some of the symptoms of naturally occurring kidney diseases. Selye concluded that he had shown that one of the principal causes of these diseases was excessive secretion of DCA, or substances similar to DCA, in response to stress. Was this conclusion justified?

THE MIXED RECEPTION OF SELYE'S RESEARCH

Most scientists at the time were convinced by Selye's research that the general adaptation syndrome was an accurate description of how animals respond to stress and had important implications for understanding human health questions. Yet many of these same scientists were skeptical about Selye's application of his findings to human health concerns.

Three factors weakened Selye's hypotheses and ultimately reduced the stature of his theories. First, scientists objected to Selye's description of the mechanism underlying the syndrome. There was virtually no evidence that there was such an entity as adaptation energy, and Selye was never able to demonstrate that people or rats ever exhausted anything as part of the stress reaction.

Second, endocrinologists had difficulty replicating Selye's results with DCA and kidney disease. Many scientists objected to the high doses and artificial conditions required before the rat developed its symptoms. Finally, research by physiologists demonstrated that DCA was a relatively unimportant hormone in the body; thus, results obtained with DCA were not physiologically relevant. This last point was simply bad luck on Selye's part, as no endocrinologist at that time knew that DCA was not physiologically active.

But the most critical factor refuting the concept of diseases of adaptation was the discovery that cortical hormones had dramatic anti-inflammatory effects, which earned Edward C. Kendall and Philip S. Hench the Nobel Prize in 1950. In his autobiography, Kendall describes the first rheumatoid arthritis patient to be given compound E (later named cortisone) as being unable to walk as a result of the debilitating pain she was suffering. Seven days later, this same woman shopped for three hours downtown and stated, "I have never felt better in my life." There was now hope for the three million people who suffered from this disease in the United States alone. While the world rejoiced, Selye's hypothesized diseases of adaptation were dealt a fatal blow.

Why was Selye held in such high regard in the 1930s and 1940s? Several factors were in his favor. During his early career, he did some excellent science. He saw a pattern of physiological response that many other scientists had seen, but was unique in identifying its importance for understanding how the body works.

Additionally, his general adaptation syndrome and the related hypothesis of diseases of adaptation very neatly tied together a hodgepodge of otherwise perplexing observations and experimental results. Selye's interpretations of the data were reasonable and insightful, based on the information available to him at that time.

☐ *EPILOGUE*

Selye's contribution to the field of endocrinology survives him and his ideas. Though the term *general adaptation syndrome* is rarely used by modern endocrinologists, studies of stress physiology and the activity of cortical hormones continue today at a much more intense pace than during Selye's time. Currently, about twenty hormones have been shown to be involved in the stress response. Small wonder, then, that administration of one hormone gave ambiguous results in experiments conducted by Selye and others who attempted to replicate his experiments. Researchers have also shown that chronic activation of the stress response is damaging to health and that people with hard-driving (Type A) personalities have a higher risk of hypertension and heart disease.

In addition to stimulating physiological research into the stress response, Selye extended his ideas outside of the traditional scope of biology. During his last decade, Selye promoted the social implications of the stress response to psychologists, sociologists, and administrators of many different types of organizations, who make stress reduction and management an important goal for economic and personal satisfaction.

QUESTIONS AND ACTIVITIES

1. What does this case show about the following aspects of doing biology?
   — tenacity in pursuing research objectives
   — the effect of background and perspective on the process of discovery
   — the importance of replicating experiments
   — consideration of alternative explanations of findings
   — the use of models in biology

2. Design an experiment that allows you to test whether the increased resistance to a stressor characteristic of the stage of resistance confers resistance to other stressors as well. Limit yourself to 100 rats. Be as specific as possible. What type of stressor would you use? What information would you need to know to determine when to subject your rats to the stressors? What would you measure?

3. Did you consider the ethics of research on animals in the previous question? What constitutes ethical treatment of animals for experiments?

4. A tremendous number of rats were being traumatized and then killed by Selye's manipulations. Can you think of any alternative designs that would test Selye's hypothesis that the triad of anatomical changes is a general response to pain or stress? More generally, in what ways (if any) should scientists consider the discomfort they cause their subjects when choosing research questions and designing experiments?

5. Propose a hypothesis for why cold-stressed rats have much shorter life spans than rats raised at normal temperatures. What types of observations or experiments could you do to test this hypothesis?

6. To test his idea of diseases of adaptation, Selye injected rats with high levels of cortical extract. Some of his rats showed some symptoms of kidney disease. Assume instead that all of his rats developed serious disease symptoms. What could you reasonably conclude from this finding? What experiment would you do next to follow up on this result?

SUGGESTED READING

Munck, A., P. M. Guyre, and N. J. Holbrook. 1984. "Physiological Functions of Glucocorticoids in Stress and Their Relation to Pharmacological Actions." *Endocrine Reviews* 5: (1): 25–44.

Sapolsky, R. M. 1993. *Why Zebras Don't Get Ulcers*. New York: W. H. Freeman.

Selye, H. 1936. "Thymus and Adrenals in the Response of the Organism to Injuries and Intoxications." *British Journal of Experimental Pathology* 17: 234–248.

Selye, H. 1946. "The General Adaptation Syndrome and the Diseases of Adaptation." *Journal of Clinical Endocrinology and Metabolism* 6: 117–230.

Selye, H. 1956. *The Stress of Life*. New York: McGraw-Hill.

CHAPTER 11

Christiaan Eijkman $\mathcal{E}$ the Cause of Beriberi

DOUGLAS ALLCHIN

☐ INTRODUCTION

In October 1886, three doctors embarked from the Netherlands on a mission of medical research that would take them almost halfway around the globe. They passed through the Suez Canal—opened only a few years earlier—and arrived a few weeks later in the Dutch East Indies (now Indonesia). Java and the surrounding islands would have fascinated them with the exotic wildlife, towering forests, and dense thickets of fibrous rattan vines, harvested by the Javanese and exported to Japan to make tatami mats. Elsewhere, trees had been cleared to grow crops brought from other tropical regions: sugar cane, coffee, cacao, and indigo. These crops meant that the Netherlands valued the East Indies as a colony.

Life for the three doctors on Java would be very different than in Europe. Western amenities were scarce. The tropical heat was everywhere. A typical Dutchman would also have to develop a taste for rice, a staple in this region of Asia.

One of the doctors, Christiaan Eijkman (pronounced "Ike-mahn," Figure 11.1), age 28, had seen the sights of Java before. He had served as an officer for the Dutch Army in Batavia. After two years, he had contracted malaria and had returned to the Netherlands. Malaria was one of many diseases common in the tropics. Cholera, influenza, dysentery, and plague were also widespread. So, too, was beriberi.

In fact, beriberi was the reason why the medical commission had come to Java. It is a debilitating disease, as indicated by the name itself. For the Javanese, the word *beri* means "weak," and doubling a word intensifies its meaning. Symptoms of beriberi include muscle weakness, weight loss, loss of feeling, and eventually, paralysis in the limbs. Fatigue can give way to confusion, depression, and irritability. In some cases fluid collects in the legs, taxing the circulatory system, enlarging the heart, and causing heart failure (see Figure 11.2). The disease can be fatal. At the time, as many as 80 percent of beriberi patients died.

In the late 1800s, epidemics of beriberi in Asia had become more frequent. Members of the Dutch government noted, in particular, that large numbers of their fleet crews and native work force were suffering. They wanted to know how to cure the dis-

FIGURE 11.1 Christiaan Eijkman. c. 1890.
Source: Koninklijk Instituut voor de Tropen
(Royal Tropical Institute) Amsterdam.

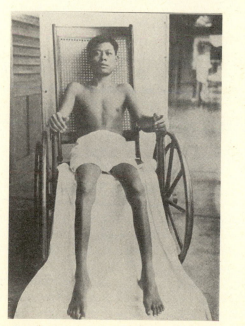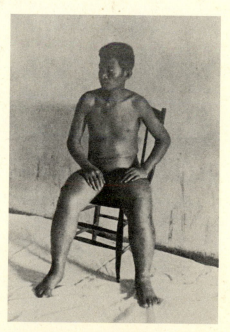

FIGURE 11.2 Beriberi patients. *Source:* (left) Edward B. Vedder, A. M., M.D., *Beriberi*, New York: William Wood, 1913. (right) Herzog, *Phillipine Journal of Science*, 1906.

ease or—better—prevent it. They sent the medical commission to find the cause of beriberi. Eijkman would eventually share a Nobel Prize for his discoveries on Java.

DISEASE, GERM THEORY, AND EIJKMAN

Beriberi was not a new disease in southern and eastern Asia. A Chinese physician had described it over 4,000 years earlier. In the East Indies, it had been reported as early as 1642. But no one knew of a cure.

Eijkman and his colleagues were not alone in trying to identify the cause of beriberi. In Japan in 1880–1881, one doctor was so swamped with beriberi patients that the hospital could not accommodate them all and they overflowed into nearby temples. What had caused the epidemic? The doctor collected data about clothing, living quarters, diet, occupation, economic status, geographical region, and season- al frequency of the disease, hoping to find clues. How would someone go about identifying the unknown cause? More generally, what can possibly cause disease?

Various researchers, both Asians and Europeans, working in Asia explained the cause of beriberi differently. Some insisted that beriberi was not a specific disease at all, but a combination of other known diseases. Others claimed it was a form of poisoning. They disagreed about which toxin was responsible, however. Was it arsenic, oxalate, carbon dioxide, or some compound produced by a microorganism? Later, some viewed beriberi as an infection, but they disagreed as to whether a protozoan, a tiny worm, or a bacterium was responsible. Another blamed moldy rice. Yet other researchers thought it was diet. But while some concluded that beriberi was due to a deficiency of fats, others implicated lack of phosphorus or proteins. For one researcher, it was insufficient nitrogen; for another, an improper *balance* of nitrogen among foods eaten. (How would someone determine which reported ideas to trust?)

The Dutch medical commission arrived with new ideas about disease from Europe. In fact, Eijkman's career reflected shifting notions of disease. Eijkman had first visited the Indies in 1885 to fulfill a contract with the military, which had helped pay for his medical education. At that time, the role of bacteria in causing disease was still a relatively new idea. When he returned to the Netherlands, however, Eijkman became excited by the expanding studies on the topic. He turned from practicing medicine to pursuing medical research. Eijkman went to Berlin to study with the world leader in the field, Robert Koch. According to Koch's "germ theory of disease," disease was the result of microscopic organisms that infected the body. Over a cen- tury later, we are well aware of "germs" and the importance of personal hygiene, community sanitation, and sterilization in medicine and food preparation. But in the late 1800s, this understanding of disease was still guiding new discoveries.

In 1880, Koch had developed a method for culturing bacteria on a solid medium instead of in a liquid nutrient broth. By spreading the bacteria across a plate, he could separate the different strains or species of a mixed culture, isolate each one, and then breed a pure culture. With this method it became much easier to isolate and identify specific disease-causing agents. In 1882 and 1883, Koch himself was able to identify the bacteria that caused tuberculosis, cholera, and diphtheria.

Outbreaks of beriberi were common in armies, navies, and prisons, all relative- ly closed communities. This was typical of infectious diseases transmitted by some "germ." Thus, when the Dutch government decided to organize a group to go to the East Indies in 1886, it sent two doctors to Germany to learn the latest techniques firsthand from Koch. Once there, they met Eijkman. Hearing about their mission, he decided to join them. In that same year, a prominent French researcher, using a method he had pioneered a few years earlier, created a vaccine for rabies. The Dutch commission took all these new methods with them to search for the bacterium that caused beriberi, isolate it, and make a vaccine. The scientists were themselves vehicles for transferring germ theory from Europe to Java.

Just over a year later, the doctors completed their work on Java. They characterized the clinical symptoms of beriberi more precisely, and reported microscopic observations of nerve degeneration in the tissues. They confirmed that a bacterium caused beriberi. But they also discovered a new infection pattern. They had not been able to infect one organism directly using another. Whereas most diseases were transmitted through a single exposure to the germ, a person had to reside in an area of beriberi infection for several weeks to contract the disease. In the case of beriberi, they reasoned, the bacterial agent had to be transferred many times. The commission returned home, leaving Eijkman on Java to direct the local medical school. He established a small laboratory, where he continued the work on beriberi. He had yet to establish a pure culture of the bacterium and, from that, develop a vaccine.

CHICKEN FEED?

Three years passed as Eijkman continued his investigations. Even using Koch's techniques, however, he was unable to isolate the beriberi bacterium in a pure culture. Then, in 1890, amid his daily activities of directing the medical school, teaching, and treating patients, Eijkman made a chance observation. He noticed that chickens in the hospital yard were suffering from conditions like those of his beriberi patients. The chickens walked unsteadily and had difficulty perching. Later they lay down on their sides—not a typical posture for chickens! (see Figure 11.3). They also had trouble breathing. He posed the obvious question: Could the chickens be infected with the same organisms that caused beriberi? Eijkman recognized the great potential for

FIGURE 11.3 Chicken with "beriberi." *Source:* from Casimir Funk, *The Vitamines,* Baltimore: Williams and Wilkins Company, 1922.

studying the disease in a population of laboratory animals where he could monitor variables more closely (Chapters 5, 10, and 14 also discuss the use of animal models). He could also change conditions experimentally. Eijkman promptly had the chickens moved to another location for further study.

When the chickens were moved, though, their health suddenly improved, with no apparent treatment! Why? Given this unexpected turn of events, what would be Eijkman's next step? Where would he look for clues?

Eijkman began to search for the factors that might help him to isolate the bacterium. He traced at least one possible difference to the chickens' new home. Previously, the chickens had been served boiled rice left over from the officers' table in the military hospital. But a new cook had given them a different variety of red rice, known locally as *beras merah*. Normally, the local rice had a reddish cuticle (or pericarp, in botanical terms). You could remove the cuticle, though, by milling or "polishing" the rice. Polished rice had a fancier white appearance and a taste that many people preferred. The cook had decided, however, that "civilian" chickens did not deserve such special white rice. So he fed them "half-polished" rice instead.

When Eijkman discovered the change, he had an important clue. The polished rice must be the source of the infection. The white, starchy portion of the rice grain must contain the bacterium for which he had searched for so long. This would certainly explain why beriberi was so prevalent in nations where rice was a staple food. Eijkman had not planned to change the chickens' diet, but neither did its effects escape his notice. The chance event revealed valuable information that he and his colleagues had missed during five years of deliberate study.

Soon, Eijkman was able to make chickens sick almost at will, simply by controlling their diet. When fed the polished, white rice, healthy chickens soon showed symptoms similar to those of human beriberi. In addition, when he fed them red rice, they became well again. They recovered as well when just the husks or cuticles of the rice—the "rice polishings"—were added to a diet of polished rice. Eijkman must surely have been impressed that, in some cases, the sick chickens regained a normal gait and the ability to fly within a few hours of eating the rice polishings. He reasoned that there must be a neutralizing agent or antidote to the bacterium in the cuticle of the rice. This could explain why healthy chickens eating red rice remained healthy, even when living in the presence of diseased birds.

Not everyone who heard of Eijkman's conclusions accepted them. Others agreed that the rice Eijkman used was responsible, but perhaps not for the reasons he specified.

PROBLEM
Identify at least two other plausible interpretations of Eijkman's findings. For one of these, design an experiment that would help confirm Eijkman's ideas and exclude the alternative explanation or vice versa.

Eijkman continued his studies while still addressing his other administrative and teaching duties. Many researchers failed to accept his conclusions because they refused to believe that the chickens' disease was the same as human beriberi. So

Eijkman characterized the disease more fully. He examined the chickens' tissues and noted the same degeneration of the nerves that the commission had identified in human beriberi. Eijkman also wanted to show the connection by transferring the disease from humans to chickens via injections of blood or other body fluids from beriberi patients—but luck failed him. Nor was he able to transfer the disease directly from one chicken to another. He speculated that the microorganism did not enter the blood itself. Instead, it might remain in the intestine, where it could produce a toxin from the starch or something else in the rice grain. The toxin, rather than the microorganism, might then enter the body and poison the nerve cells. Eijkman would have to modify his investigations.

OF RICE AND MEN

Five more years passed as Eijkman continued to search for the beriberi bacterium and the toxins it produced. Meanwhile, two researchers (one Japanese, one French) independently isolated the bacterium that caused bubonic plague. And in India, over 45,000 people received a new cholera vaccine. Compared to those not inoculated, 70 percent fewer died. Germ theory was still controversial, however. In 1892, a skeptic in Germany swallowed a vial of live cholera bacteria to demonstrate his belief that the bacteria did not cause the deadly disease. Indeed, he did not get sick.

Eijkman had still not demonstrated conclusively how polished rice was part of the process by which bacteria caused beriberi in humans. He needed a properly controlled experiment. Of course, he might well have decided to feed several individuals nothing but polished rice, as he had done with chickens. (Would this have been ethical?) Eijkman turned instead to institutions. There, diets would already be determined. The large number of cases would also help ensure that the results would not be due to chance or mere coincidence in a small group. He persuaded the prison at Tolong, where 5.8 percent of the population suffered from beriberi, to substitute undermilled, or half-polished, rice for white rice. All cases of beriberi were cured. But, as Eijkman noted later, this merely confirmed the potential effectiveness of the cure. It did not demonstrate that a bacterium in the polished rice had initially caused the disease. This would require comparing individuals who consumed the different types of rice.

Eijkman thus enlisted A. G. Vorderman, supervisor of the Civil Health Department of Java, to help survey the incidence of beriberi on a wide scale. In each prison on Java, prisoners ate either polished rice or half-polished rice, according to local customs. In some cases, prisons served a mixture. Here was a **natural experiment**, a case where the desired experimental controls existed without manipulation by the investigators. Fortuitously for Eijkman and Vorderman, the experiment was already in progress. Between May and September of 1896, Vorderman led an exhaustive study of beriberi in 100 prisons on Java and a small neighboring island—a survey that embraced nearly 280,000 prisoners. He reported the distribution of beriberi in the 100 prisons and its frequency among prisoners as follows:

	Number of Prisons	Number with Beriberi	Percentage of Prisons with Beriberi	Frequency among Prisoners
Half-polished rice	35	1	2.7%	1 in 10,000
Mixture	13	6	46.1%	1 in 416
Polished rice	51	36	70.6%	1 in 39

Vorderman also considered other possible sources of the beriberi bacterium, focusing especially on hygienic factors (why was this comparison important?):

	Number of Prisons	Prisons where Beriberi Found	Percentage of Prisons with Beriberi
Age of buildings			
40–100 years	26	13	50.0%
21–40 years	32	11	34.4%
2–20 years	42	19	45.2%
Floors			
Impermeable	58	24	41.4%
Partly permeable	13	7	53.9%
Permeable	29	12	41.4%
Ventilation			
Good	68	28	41.2%
Medium	11	8	72.7%
Faulty	21	7	33.3%
Population density			
Sparsely populated	73	32	44.6%
Medium population	1	1	
Overcrowded	26	9	34.6%

Many diseases were more prevalent among those on lower ground. Vorderman also collected data indicating that beriberi did not correlate with lower altitude. Nor did the incidence of other diseases match that of beriberi. In four prisons, Vorderman noted further, the number of cases of beriberi increased with the arrival of a prisoner who already had beriberi.

PROBLEM
What conclusions can be drawn from Vorderman's study beyond what Eijkman could conclude from his study with chickens? Reconsider your earlier assessments. How do Vorderman's results support Eijkman's explanation and/or other alternatives?

As a brief aside, consider these studies and the Dutch effort to cure beriberi from a Javanese perspective. First, why were so many prisons available for scientific study? The Dutch were managing over a quarter of a million prisoners on one island! In the late 1800s, Java was one of the most densely populated areas in the world, with between 30 and 35 million inhabitants. Still, almost 1 percent of the population was in prison. From the local perspective, the Dutch colonials were invading foreigners. The prisons, all military prisons, reflected how the Dutch dealt with the Javanese opposition to their occupation. Vorderman's survey took advantage of that exercise of colonial power.

Second, though more Javanese than Dutch suffered from the disease, the Dutch colonials had more at stake than simply aiding the indigenous population. The disease took its toll on the local work force. Beriberi interfered with the Dutch "trade" in the region. The Dutch thus valued a cure for economic reasons. Likewise, no one had offered the Javanese the tools or resources to study the disease on their own. Though Eijkman and Vorderman addressed fundamental biological questions, their research on this occasion was also motivated by the Dutch economic interests and facilitated by its military presence. Social factors mixed with "pure" science.

BERIBERI AFTER EIJKMAN

Eijkman left Java just as his collaboration with Vorderman was ending—again due to illness. Back in the Netherlands, he briefly continued his studies on beriberi. Unsuccessful in his efforts to isolate the bacterium, he focused on the cure instead. He showed that water and alcohol extracts of the rice cuticle could cure the disease as effectively as the polishings themselves. He confirmed that the curative factor was destroyed when heated over 120° C. It could also pass through a membrane, such as the cell membranes of an intestine. Eijkman published his results and then turned to other research inspired by his visits to the tropics, leaving others to pursue the remaining mysteries of beriberi.

Beriberi was important enough that research had been occurring in several places besides Java. There were major efforts in Japan, Malaya, and the Philippine Islands. (In Japan's war with Russia in 1904–1905, 4,000 soldiers died of beriberi.) Eijkman's and Vorderman's results were dramatic. But consider how others would have heard about them. If you were studying beriberi in southeast Asia in the 1890s, how would you know that someone else was also studying the same disease in a nearby region? If you were aware of such work, how would you find out about the results? What about differences in language? With no formal network or institution for communicating findings, it is difficult to share ideas, build on the work of others, or benefit from criticism. In this case, there was no scientific community. Research was fragmented, and research news traveled slowly. In fact, most scientific studies sponsored by colonial powers during this period were published in Europe. Eijkman and Vorderman followed this pattern—writing in Dutch, no less.

Nevertheless, in the early 1900s research began to focus more on rice in the diet. Large-scale studies like Vorderman's continued through 1912, in each case confirming the findings on rice. Between 1905 and 1910, major institutions—armies, navies, prisons, insane asylums, and leper colonies—finally began to change their primarily white-rice diets. Many researchers also continued to search for the bacterium or toxin in the rice and the identity of the curative factor in the rice cuticle.

In Java, another Dutch doctor, Gerrit Grijns, succeeded Eijkman at his laboratory. Grijns disagreed with how Eijkman had interpreted his results, however. For Grijns, it was not the rice that was toxic, nor the polishings that effected a "cure." Rather, something essential seemed to be missing from the rice once it was polished. The rice cuticle must have contained some critical nutrient. In other words, Grijns saw beriberi as a nutrient deficiency, not the result of some "germ."

PROBLEM
How would Grijns have explained Eijkman's and Vorderman's data?

Pursuing his alternative hypothesis, Grijns examined the ability of other starchy foods to produce "beriberi" in chickens. Diets of either tapioca root or sago (the starchy pith of a palm) could also produce the disease. If there was a microorganism or toxin, it was certainly not unique to rice. Grijns also looked for other sources of the curative or missing factor. He tested each one by adding it to a chicken's diet of polished white rice. He found that several beans, notably the mongo bean, known locally as *kachang-ijo*, could "cure" or prevent beriberi. Grijns's results dramatically undermined and virtually reversed Eijkman's conclusions. Beriberi patients did not suffer from some disease-causing agent in their diet. Rather, they suffered from some health-related element missing from it.

The work on beriberi by medical researchers eventually intersected with independent investigations on nutrition by biochemists in Europe. In England in 1910–1912, one researcher (Frederick Gowland Hopkins) fed young rats highly purified forms of the basic ingredients known to be essential for any diet: proteins, fats, carbohydrates, water, and salts. Though apparently fully nourished, the mice ceased to grow. When given as little as 2 or 3 cubic centimeters of milk per day, they began to grow again. Such amounts were insignificant in terms of their protein or energy. The researcher concluded that "accessory factors" in the milk were necessary, though only in extremely small amounts.

During the same period, several individuals working independently—Casimir Funk, a Pole working in London, E. S. Edie, also in England, and Umetaro Suzuki in Japan—each isolated an anti-beriberi chemical. They recognized more clearly how beriberi and similar diseases were linked to dietary requirements. Scurvy and pellagra, along with beriberi, were all deficiency diseases. That is, they resulted from something essential missing from the diet. Because the vital elements included substantial nitrogen, Casimir Funk called them "vitamines." Later, the specific factors were labeled as we now know them: vitamin C was associated with scurvy; vitamin B_1, with beriberi; niacin (also in the B complex), with pellagra; and vitamin D, with rickets. Ironically, Eijkman did not accept these conclusions when they were first introduced. The significance of his work in furthering the study of vitamins was acknowledged, nevertheless, by a Nobel Prize in 1929, awarded jointly to one of the biochemists (Hopkins) and to Eijkman, then age 81.

☐ EPILOGUE

The potentially fatal effects of deficiency diseases demonstrate vividly that vitamins are important in our diet. This may not seem obvious given the small amounts that we need. Just how small these amounts were became clear when the "beriberi vitamin," thiamine, was isolated in 1925. From 300 kilograms of rice polishings, a pair of Dutchmen (again working in Java) extracted only 100 milligrams of thiamine. Even in the rice cuticle, which could prevent beriberi, the vitamin was present in only a few parts per million. Vitamins are not ordinary nutrients.

Biochemically, vitamin B_1 serves as a coenzyme, meaning that it participates in an enzymatic reaction but does not itself catalyze the reaction. Only small amounts are needed because each molecule is used many times. One specific role for thiamine is in the energy pathway between glycolysis and the Krebs cycle (Chapter 7). When the 3-carbon product of glycolysis is split, it releases energy and yields a 2-carbon fragment. This fragment is temporarily attached to a modified thiamine, which (as a coenzyme) transfers it to the first enzyme in the Krebs cycle (see the missing stage in Figure 7.3(B)). Thiamine is therefore critical to virtually every cell in the body. Nerve cells, which use much energy, are strongly affected by its absence, accounting for the clinical symptoms of the disease.

Abundant thiamine is found in the cuticle and bran husk of cereals. Thus today's whole-grain breakfast cereals and breads tout their image as so-called "health foods" in part because they contain the essential B_1 vitamin. Thiamine is present as well in yeast, legumes, and green leafy vegetables. (The thiamine in meats is lost when they are cooked.) Many foods, such as enriched white flour, are supplemented with thiamine. Otherwise, it might be missing from the highly processed diets of many people in industrialized countries.

Why had beriberi suddenly become more prevalent in the early 1870s? During that period, Westerners introduced steam-driven mills to the East. The mills replaced more traditional methods of hand-pounding rice (Figure 11.4). The highly effective

FIGURE 11.4 (A) Steam-powered mill for removing bran and cuticle from rice. (B) Traditional method for hand-pounding rice. The frequency of beriberi increased with the introduction of steam mills. *Source:* Edward B. Vedder, A.M., M.D., *Beriberi*, New York: William Wood, 1913.

milling process stripped the essential vitamins from the rice with increased efficiency. As steam-milled white rice became more common, so too did the occurrence of beriberi. From different perspectives—biochemical, dietary, and social—what had been the cause, or causes, of beriberi?

QUESTIONS AND ACTIVITIES

1. What does this case show about the following aspects of doing biology?
    — chance or accident
    — theoretical perspectives in interpreting data
    — the distinction between causation and correlation
    — growth of knowledge through small cumulative additions versus through major conceptual reinterpretations
    — the role of individuals versus groups in making a discovery
    — scientific communication and communities of researchers
    — the cultural and economic contexts of science

2. Who discovered vitamins? When? Consider both the contributions and the historical perspectives of Eijkman, Vorderman, Grijns, and others (including Hopkins, Funk, Edie, and Suzuki). Discuss what it means to make a discovery in science. How would you have advised the Nobel Prize committee giving the award on this occasion?

3. Discuss how Grijns's interpretation of Vorderman's data reflects both the importance and the limits of a controlled experiment. How were Vorderman's conclusions significant? Describe how his conclusions could also have been mistaken. (If you have read Chapter 1, 4, or 10, you might relate your comments to other cases of correlation and causation.)

4. A doctor in the Japanese Navy made a number of observations in the early 1880s trying to identify common factors among disease victims. He found that:
    a. Cases of beriberi were most frequent from the end of spring into summer but were not isolated to those seasons.
    b. The frequency of disease also varied considerably from one ship to another and from one station to another within a ship.
    c. Upper-class individuals suffered less than sailors, soldiers, policemen, students, and shop boys.
    d. The disease was more prevalent in large cities, but even people living in the same area did not suffer equally.
    How might Eijkman have later explained these observations?

5. Between 1885 and 1906, 17 different researchers claimed to have found the microorganism that caused beriberi. Other researchers, including Koch, had searched for the infectious agent and failed to find one. They concluded that beriberi was not bacterial at all. From the perspective of someone who thought that beriberi was infectious, suggest several reasons why Eijkman, Koch, and others might have failed. Was failure to find a pathogen definitive in this case? Where should the burden of proof lie here?

6. Consider the various claims about beriberi, microorganisms, and diet in 1900.
   a. If you were a researcher at this time, what reasons would you have to investigate infection versus diet as a cause of beriberi?
   b. If you were a public administrator in Java in 1900 with a limited budget, what programs would you support to control the incidence of beriberi? Should you inform the public about consumption of half-polished rice, improve sanitation of rice storage and transport, wait, or do something else? How would you respond to potential critics?

   Based on your responses, consider how scientific uncertainty affects different types of decision making.

7. Based on Eijkman's work, two researchers took a healthy work force to a previously isolated area of Javanese forest in 1906. They fed one-half of the workers white rice and the others, a more complete diet. They continued until the workers who were fed only rice became ill with beriberi. They then switched diets between the two groups. The first group was cured, while the second group became ill. Comment on the design of this experiment, both in terms of experimental controls and ethics.

8. Consider the causes of beriberi on different levels. What was the cause of beriberi on a biochemical level? On a dietary level? On a social, cultural, or economic level? How is each perspective associated with an alternative solution for reducing the frequency of beriberi?

SUGGESTED READING

Funk, C. 1912. "The Etiology of the Deficiency Diseases." *Journal of State Medicine* 20: 341–368.

Vedder, E. B. 1913. *Beriberi*. New York: W. Wood.

Williams, R. 1961. *Toward the Conquest of Beriberi*. Cambridge, MA: Harvard University Press.

CHAPTER 12

Western Science, Pain, *&* Acupuncture

DOUGLAS ALLCHIN

☐ *INTRODUCTION*

In 1971, the United States was still basking in its recent success of landing the first man on the moon—perhaps the greatest technological feat of human history. An electronic microprocessor—a computer "chip"—had just been introduced. Texas Instruments had begun marketing the first pocket calculator: it could add, subtract, multiply, and divide—and it sold for $150! America was still experiencing "hippies" and their youthful counterculture of exploring psychedelic drugs and heroin. Much fanfare accompanied President Richard Nixon's effort to reestablish U.S. ties with the People's Republic of China. For several decades, relations with its communist regime had been severed. China was the most populous nation on the globe, with one-quarter of the earth's people crowded within its borders. It had a rich tradition of culture and technological discovery dating back thousands of years, but was also largely a nation of peasants.

Imagine yourself as an American physician visiting a Chinese hospital in 1971. Some doctors there tell you that they can alleviate pain by inserting needles at specific points in your skin and then twirling them slowly. The points may be quite remote from the source of pain. Would you believe them?

Next, you witness a surgical operation where the main source of pain control is from a needle inserted in the patient's forearm. The surgeon makes a 14-inch incision around the left side of the thorax, cuts two ribs, and removes a lobe of a tuberculosis-infected lung. Meanwhile, the patient—still conscious—chats with the surgeon (Figure 12.1). At the end of the two-hour procedure, the patient sits up and leaves the operating table under his own power.

Finally, the doctor, who knows some Western medicine, explains that it is all quite reasonable if you understand traditional Chinese medicine and its basis in the flow of the life force and the harmony of nature. How would you respond?

The use of needling as a form of medicine—a practice known as **acupuncture**—extends back more than 2,000 years in China (Figure 12.2). Acupuncture received widespread exposure in the United States, however, only in the wake of

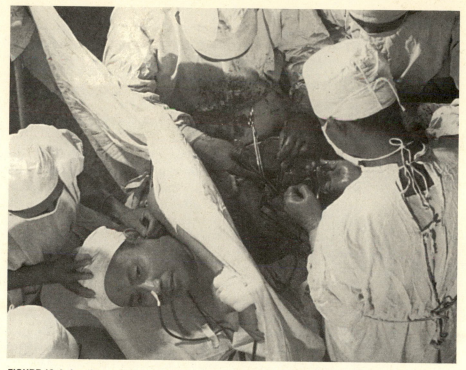

FIGURE 12.1 A patient remains conscious during surgery with acupuncture analgesia. *Source:* Eastfoto.

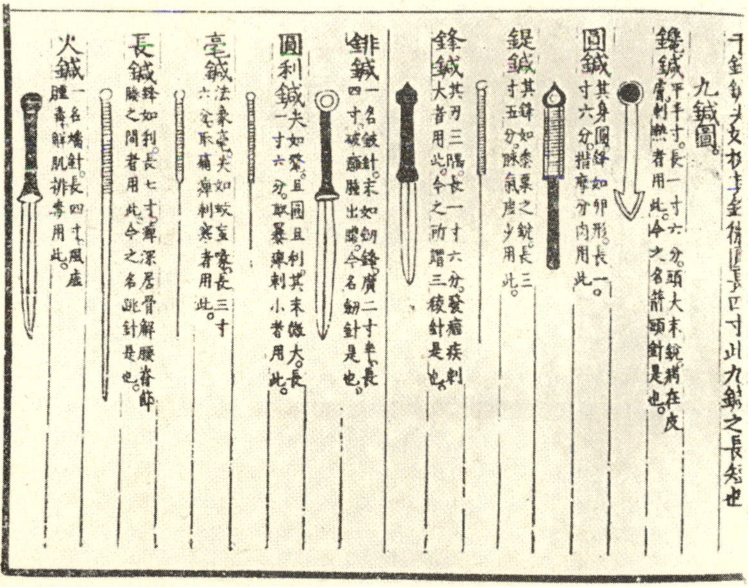

FIGURE 12.2 Different types of acupuncture needles, as depicted in a 1601 Chinese text. *Source:* from *Celestial Lancets: History and Rationale of Acupuncture and Moxa* by Gwei-Djen Lu and Joseph Needham. Reprinted with the permission of Cambridge University Press.

Nixon's diplomacy. American physicians could not explain how acupuncture could suppress pain. It followed no known physiological mechanisms. Indeed, it seemed contrary to scientific and commonsense notions about pain. What would have been an appropriate response in 1971? Could acupuncture be interpreted or explained in terms of Western science? Was it even worthy of serious scientific attention?

A SKEPTICAL RESPONSE

The spectacle of surgery done under acupuncture startled many Western physicians. Chinese doctors also used acupuncture to alleviate other sorts of pain, such as headaches, toothaches, and chronic pain in joints and muscles. They inserted needles at different points specific for each treatment, as prescribed by the experience of many centuries. The Chinese even used needling to treat other ailments: hiccups, insomnia, asthma, blindness, ulcers, vitamin E deficiency, and (more recently) drug and smoking addictions. That was quite an extraordinary list for such a modest procedure.

It is not surprising, then, that some American doctors doubted the Chinese claims. For them, acupuncture was unscientific. They alleged fraud and warned of possible abuse by "quackupuncturists." In fact, many acupuncture clinics that opened in the United States—and some that had existed previously in various Chinatowns in major U.S. cities—were soon closed by government authorities. But as Western physician after Western physician offered eyewitness testimony and as doctors began to replicate acupuncture's control of pain in U.S. hospitals, the claims of fraud gradually became untenable. Public decisions based on early judgments about acupuncture's scientific validity, however, often remained unchanged.

Many people acknowledged that acupuncture helped patients, but they were nonetheless unimpressed. They gave two reasons. First, a patient sometimes recovers from his or her condition even without treatment. For these cases, it would be inappropriate to credit acupuncture. To assess the effect of the needling alone, they argued, you must conduct a controlled study in which you compare patients treated with acupuncture with those who receive no treatment.

From a traditional Chinese perspective, however, this idea posed an ethical problem: Why would you withhold a treatment you knew to be effective? Chinese medicine stresses the result for the patient. Research is secondary. Therefore, you do not refrain from treating someone just for the sake of a test. Basically, for the traditional Chinese, if you already knew *how* to help a patient recover, you did not also need to know *why* the procedure worked, especially if your research might be at the cost of a patient's well-being.

A second reason for disregarding acupuncture, according to some critics, was that pain might be suppressed merely through psychological suggestion, not by the needling itself. In other words, pain control might have resulted from some "unscientific" influence, such as hypnosis, that was not worthy of serious medical attention.

Such psychological effects were already well known from drug studies. Even biologically inactive substances—known as **placebos**—may sometimes have a positive effect. Physicians have accepted for several decades that as many as one-third of all patients respond favorably, even when treated with a placebo or "non"-treat-

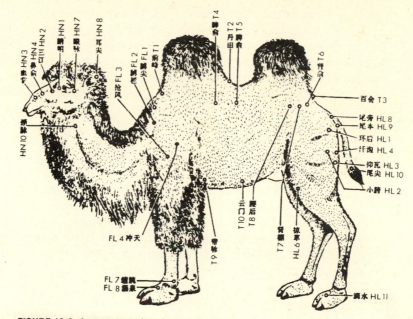

FIGURE 12.3 Acupuncture chart for a camel. Diagrams of needling points for animals first appeared in China centuries ago. *Source:* Camel diagram from *Veterinary Acupuncture* by Alan M. Klide and Shiu H. Kung. Copyright © 1977. Reprinted by permission of University of Pennsylvania Press.

ment, as long as they believe that they are receiving a real treatment. Many Westerners were especially suspicious when they discovered that the Chinese tended to screen patients: not all were deemed eligible for acupuncture. At the same time, Chinese doctors did not distinguish so sharply between psychological and physiological mechanisms. The treatment itself was effective. A Chinese doctor might have replied: "Is psychosomatic healing not healing?" Chinese and Western standards for "good" medicine and science emphasized different values in this instance.

Of course, you could test for the effect of suggestion. For example, you could check acupuncture on a person or organism who can feel pain but is not susceptible to suggestion. In fact, infants respond to acupuncture. So do animals. The Chinese have treated farm animals—horses, pigs, water buffaloes, etc.—with acupuncture since at least the Yuan period (1300s—before the European Renaissance). They have produced diagrams of the specific acupuncture points on animals—even camels! (Figure 12.3) Since the 1970s, numerous veterinarians in the United States have adopted acupuncture as part of their practice.

Another possible approach to testing suggestibility—similar to methods used in drug trials—is to prevent the patient from knowing whether he or she is receiving treatment. At the same time, you also need to guard against the possible suggestive influence (however subtle or indirect) of a doctor who knows which patients are receiving treatment and which are not. Such trials would be **double-blind**: both patient and doctor are "blind" to who is in the control or test group.

The nature of acupuncture poses special problems for double-blind tests. First, it is difficult for patients not to know whether they are receiving acupuncture. One solution has been to use "sham acupuncture," needling at nonacupuncture points. Even in this case, however, the specific points must be identified by a trained acupuncturist who knows whether each point is authentic or sham. Double-blind tests are also difficult because the acupuncturist depends in part on how the patient responds to the insertion of the needle to know whether it is placed precisely. There are no overt anatomical indications of the points, but a patient sometimes feels a slight distension or numbness around the point when the needle is properly inserted. The procedure should be interactive, making mutual ignorance difficult.

Studies proceeded with these qualifications in mind. Some examined acute pain (sharp, brief pain from such sources as intense heat or sharp objects), while others examined chronic, or long-lasting pain (such as backaches or arthritis). When acute pain stimuli were given to humans, mice, cats, horses, rats, and rabbits, needling of true points clearly suppressed pain, whereas needling of sham points produced very weak effects. By contrast, the results from clinical studies of chronic pain were more complex. Several independent studies since 1971, each based on the patients' own evaluations of pain, are summarized in the following table (the differences between groups are statistically significant):

	Effectiveness (reported cases of pain relief)
Sham acupuncture	33%–50%
Placebos of needles taped to the skin	30%–35%
"True" acupuncture	55%–85%

By comparison, morphine, accepted in the West as the most widely effective painkiller, relieves pain about 70 percent of the time. Also, recall that approximately one-third of all patients respond positively to a placebo for almost any medical treatment.

PROBLEM
Based on these results, would you conclude that acupuncture works physiologically or psychologically? Explain. How do you interpret the difference in results for acute versus chronic pain?

CHALLENGE
Which treatment would you select as the experimental control in further studies: sham acupuncture or needles taped to the skin? Why?

AN EXPERIMENTALIST RESPONSE

While some medical researchers regarded acupuncture as so much hocus-pocus, others were curious to know how acupuncture might work. For answers, though, they had to commit themselves to investigating a phenomenon whose exact nature in Western terms was still very much uncertain. Indeed, in the early 1970s, knowledge of how we perceive pain was itself quite incomplete.

Some researchers applied the biological principle that function relates to structure and supposed that acupuncture points might have some special anatomical or physiological properties. For example, might acupuncture needles stimulate some particular kind of nerve ending or bundle of nerves? Despite studies of cells and tissues in the skin, though, no structure was found that was unique to these points.

Other studies looked at the electrical properties of the skin. Initially, acupuncture points seemed to be areas of low skin resistance. Veterinary acupuncturists now frequently use electrical devices to help them locate points. At the same time, some experts now note how difficult it is to measure skin resistance and suggest that many studies are not reliable. There does seem to be lower skin resistance at acupuncture points, but this has not yet helped explain how acupuncture works.

By contrast, there was relatively quick confirmation that acupuncture for acute pain stimulates one particular kind of nerve. Most painful stimuli are carried along small fibers. Acupuncture, though, apparently stimulates larger fibers (type II and III muscle afferents). One recent theory suggested how the pain and acupuncture impulses might interact. The interaction was built on a mechanical analogy. According to the theory, there was a figurative "gate" in the spinal cord where the two types of fibers converged. Only one impulse could be conveyed to the brain. The large fiber would synapse with the small fibers, inhibit them, and prevent further impulses. This "gate-control" theory could thus explain how gentle needling might "switch" off perceptions of pain—at least where nerves entered the same segment of the spinal cord.

Other researchers noticed that the optimal effects of acupuncture often occur after several minutes—too slowly to be explained by nerve impulses. They wondered if the needling might release some factor in the blood. If so, they reasoned, the blood from one organism subjected to acupuncture, when transfused into a different organism, should have a measurable effect. One team set up a system of cross-circulation between pairs of rabbits by cross-linking the veins in their legs. Indeed, the acupuncture on one rabbit allowed the other to withstand stronger painful stimuli. Cross-injections of cerebral-spinal fluid also worked. Acupuncture seemed to trigger the release of an unknown hormone or similar "messenger" substance.

The puzzles of acupuncture in 1971 intersected with findings from other research on pain relief, already under way. Morphine, a derivative of the opium poppy, had been used medically for some time to relieve extreme pain. It was addictive, however, which limited its use. No one knew how it worked. In the 1960s, due to a widespread counterculture movement, the use of heroin and other morphine-related opiates to explore altered states of consciousness increased dramatically. So, too, did the number of cases of addiction. Due to the social consequences, funding became available for research—and some unexpected facts emerged.

In 1973, researchers discovered that opiate molecules fit into receptors on the surface of cells in the brain. They confronted a disturbing question: Why would substances that were not part of normal human physiology have such functional "slots" in human cells? Researchers inferred, somewhat reluctantly, that there must be some unknown natural "opiate" that controlled pain in the body. After a brief search, several such substances were found. Among them was **endorphin**, released from the pituitary gland into the bloodstream. The curious effect of acupuncture through the blood now had a possible explanation that could be investigated further.

The study of endorphin's effects was facilitated by a chemical, *naloxone*. Researchers determined independently that naloxone inhibited or interfered with the effect of both natural and artificial opiates by blocking their receptors. In 1976, one study addressed the effects of naloxone injected just prior to acupuncture in rabbits:

Treatment		Pain Relief?
Acupuncture		Yes
Acupuncture + naloxone		No
Acupuncture + saline*		Yes
naloxone	(no acupuncture)	No
saline*	(no acupuncture)	No
Acupuncture at nonpoints	(sham acupuncture)	No
Handling, restraint, and pain-testing	(no acupuncture)	No

*"Saline" represents an injection of a physiologically inert solution of salts.

PROBLEM
Do these data effectively support the notion that acupuncture works by stimulating the release of endorphins? What is the role of each treatment in reaching a conclusion in this study? (For example, identify pairs of treatments that you compare with each other.)

CHALLENGE
Describe how you might further apply this procedure to test for suggestibility in humans.

POINTS EAST AND WEST

The notion that an apparently painful stimulus might reduce pain was paradoxical enough. But even more puzzling for Westerners were the patterns of needling. As noted earlier, Chinese doctors do not insert acupuncture needles haphazardly. There are specific points. Sometimes, the points are quite remote from the site of their intended effect. Thus, you would insert a needle between the thumb and forefinger (a well-known point called *ho-ku*) to treat either a headache or abdominal cramps! For coughing or a fever, you would use a point above the third toe. For Westerners, at least, the correlations made no anatomical sense.

The Chinese explanation for acupuncture, however, accounted for why the points and their effects could sometimes be so distant from each other. The Chinese conceived the body in a wholly different way. According to their traditional theories, the body is fundamentally maintained by a life force, **qi** (pronounced as a short, breathy "chee"). *Qi* flows through the body along several intersecting meridians, or channels (Figure 12.4). There are 12 primary meridians, each corresponding to a

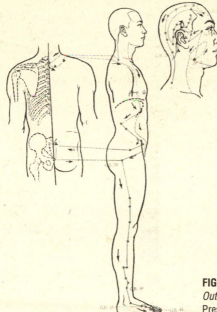
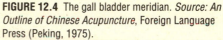

FIGURE 12.4 The gall bladder meridian. *Source: An Outline of Chinese Acupuncture*, Foreign Language Press (Peking, 1975).

major organ (liver, stomach, spleen, gall bladder, etc.). They also correspond to the 12 yearly cycles of the moon. The flow of *qi* along the meridians maintains a balance between *yin* and *yang*, the two complementary forces of the universe, according to Chinese philosophy. Thus, some meridians or channels are yin, others yang.

The flow of *qi* along the meridians is part of the traditional Chinese explanations for health and illness. When the flow is impeded or imbalanced, disease, malfunction, or pain results. To restore the balance, needles are inserted at points along the appropriate meridian. The needles either promote or impede the flow of *qi*, reestablishing the balance of yin and yang. Because the *qi* flows along the meridians, a needle inserted at one point may affect a distant organ along the same meridian. Thus a needle at *ho-ku* can affect a headache or abdominal cramps because all lie on the large intestine meridian. In the Chinese view, part of the acupuncturist's skill is diagnosing which meridians are affected and where along those meridians needles should be placed.

In many ways, the Chinese concepts resonate strongly with W. B. Cannon's notion of homeostasis (Chapter 9). The concepts of yin and yang, especially, seem to parallel the sympathetic and parasympathetic nervous systems in their balanced interaction. In both cases, an imbalance is associated with illness, and restoring the balance is associated with health and well-being.

Historically, acupuncturists also had to determine exactly where the points were and how they were connected along meridians. No one could predict, apparently, the sometimes zig-zagging pathways of the meridians—and there is no reason in the Chinese view why they follow the paths as now described. Instead, the meridian maps represent the collected experience of generations of acupuncturists.

The concepts of *qi* and meridians can be problematic from a Western perspective because no anatomical structures define the meridians, and no measurable force can be identified as *qi*. Chinese and Western traditions have different, even incompatible "geographies" of the body. A Westerner may be inclined to think that the notions of meridians and *qi* are superfluous—perhaps relics of an ancient and discredited cosmology—but they are essential from the perspective of actual practice. Acupuncturists use the meridian maps to assess where needles should be placed (though they also learn specific points traditionally effective for common conditions).

Acupuncturists also use the notion of *qi* in inserting the needle. As noted earlier, the patient can feel when needles are placed in the correct location. The sensation is known as *de qi*, or "striking the *qi*," reflecting the view that the patient perceives how the flow of *qi* changes. Patients can sometimes also feel the numbness of *de qi* spread along the line of the meridian. For the practicing acupuncturist, at least, the traditional theory cannot be easily discounted.

Westerners could not explain the relationship of points in acupuncture. Not that they hadn't noticed similar phenomena. They were familiar, for example, with *referred pain*. In these cases, pain from an injured internal organ is felt on the surface of the body, but not always above the organ. Perceptions occurred within the same segment of the body, however (based on where nerves branch from the spinal cord). The sometimes distant separation of acupuncture points and their effects was still puzzling in terms of referred pain.

Westerners were also familiar with another set of points, discovered at the end of the last century, now known as *trigger points*. They evoke pain when pressure is applied. According to one physician, additional pressure on these points can also alleviate pain. Researchers found in 1971 that there is a strong correlation between the location of the trigger points of the West and the acupuncture points of traditional Chinese medicine. But again, no one knows why trigger points produce pain, sometimes at specific locations remote from the point. In neither case—for trigger points or referred pain—is there an explanation in Western terms why these points might be related to pain relief.

PROBLEM

In recent years, many acupuncture points have been found that do not lie on traditional meridians. How would you expect a Western skeptic to interpret this fact? How might a Chinese doctor interpret the same fact? Explain how each interpretation might reflect existing theoretical perspectives.

The contrast between the apparent effectiveness of acupuncture and explanations that did not fit Western conceptions posed several dilemmas for U.S. doctors and medical researchers. Should they reject the Chinese claims as unscientific because they found the explanations unacceptable? Could they reject the explanations while offering no viable alternative? Could Americans split the demonstrated effectiveness and the explanations and accept one but not the other, while the Chinese saw them as inseparable in practice? Even if Westerners developed their own interpretation of acupuncture, could they disregard the Chinese explanations? It was difficult to dismiss the Chinese explanations out of hand, despite their apparent implausibility in terms of Western physiology, because the Chinese had discovered and developed acupuncture over several centuries, whereas Westerners had not.

☐ *EPILOGUE*

The introduction of acupuncture to the United States in the 1970s illustrates how different cultural traditions in science sometimes converge. Both scientific knowledge and attitudes about experimentation and methods of investigation differed. The different cultural contexts also guided subsequent approaches to acupuncture and to medicine more generally.

The Chinese have largely accepted Western medicine, though not exclusively. Western and traditional Chinese systems exist side by side. One American acupuncturist trained in China noted, "In China, the idea is: they do not only give you Chinese traditional medicine, because if you only learn Chinese traditional medicine, your mind will get stuck." Acupuncture is valued in China, particularly in an economic context. It is a relatively "low-tech" form of medicine. It requires little equipment, though it does require expertise and substantial training for the acupuncturist. It is a labor-intensive rather than capital-intensive form of medicine.

By contrast, acupuncture is still widely viewed in the United States as an alternative or "folk" medicine. It has peripheral status. Many insurance companies or health plans do not pay for acupuncture treatments. Even Western doctors sympathetic to acupuncture often recommend it only when Western medicine fails or is *first* shown to be ineffective. Many practitioners and health administrators say that scientific assessments leave the effectiveness and explanations of acupuncture still uncertain. They also cite the potential for fraud.

What we know about acupuncture has been and will be shaped by the research that is done. Because acupuncture involves no product to sell, drug companies and other investors have had little incentive to fund acupuncture research. There is no opportunity for profit. Research on endorphinlike molecules that may relieve pain, on the other hand, has been well funded. The prospect for knowing more about acupuncture thus depends on certain sources of funding to support research.

Research on pain and acupuncture is far from complete, but the picture is becoming clearer. The system of interactions appears to be quite complex. The "gate-control" theory, for example, now seems far too simple. There are numerous nerves that originate in the brain and end at more peripheral points. There, they inhibit the transmission of impulses towards the brain. Acupuncture activates many of these inhibitory systems at several levels. In quite different studies, acupuncture has also been linked to increased levels of cortisol, a steroid hormone released from the adrenal cortex (see Chapter 10). This suggests a link to immune responses. If confirmed, these might help explain acupuncture's other reported health effects in Western terms.

Cross-cultural perspectives in medical science have become increasingly important recently. Many researchers are interested in medical treatments among cultures in nonindustrialized nations. Native practitioners often use plants that are not familiar to Westerners. Many drug companies are investing heavily to research whether indigenous herbal treatments can cure various diseases. Where effective, they want to look for their "active" chemical ingredients. By contrast, they are usually not interested in the context of indigenous medicine, the explanations for various cures, or how they were discovered. In what ways do you think we might learn from the "science" in these other cultures?

QUESTIONS AND ACTIVITIES

1. What does this case show about the following aspects of doing biology?
   — controlled experiments
   — double-blind studies
   — theoretical perspectives in interpreting experimental results
   — the cultural context of the norms of scientific practice
   — the burden of proof and the role of skepticism
   — acceptance versus pursuit (research) of explanations

2. What level of skepticism or acceptance do you think was initially warranted in this case? Given both the long history of acupuncture in China *and* the lack of controlled studies, where did the burden of proof lie in 1971? How would these judgments affect research funding as well as public policy?

3. What do you think may have motivated such strong criticism and concerns about fraud in this case? Were such initial criticisms valuable to developing scientific knowledge or not? Socially or institutionally, how might you either encourage or suppress such motivations or otherwise regulate their effect?

4. The effectiveness of acupuncture on drug addiction, asthma, hypertension, and other conditions is still being investigated. Do you think these claims need to be scrutinized with the same vigor as acupuncture's control of pain before these treatments are adopted in the United States? Why or why not?

5. Suggest ways you might design a double-blind test for assessing the effectiveness of acupuncture. Suggest other possible ways to investigate whether the effects of acupuncture are produced by psychological suggestion.

6. In this case, Western scientists generally trusted claims by U.S. doctors more than those by Chinese physicians. Suggest several reasons, both justified and unjustified, why this might have been so. Especially given the apparent implausibility of the Chinese claims in the United States, suggest how a Chinese physician—credible among Chinese peers—might have established credibility in a community of Western scientists. Who qualifies as an expert for evaluating the claims of acupuncture?

7. Some specific studies reported pain relief from acupuncture, while others found no significant relief. How do you reconcile different studies that give both positive and negative results? For example, does one unsuccessful study undermine a string of apparently successful reports? By contrast, how might one criticize a study that fails to produce an effect? More generally, what makes an experiment decisive? How do you assess where the burden of proof lies?

8. The "discovery" of acupuncture by the West is not unlike other discoveries in science that can introduce new technologies or medical practices into society. Describe a system to address such new practices whose effectiveness is deemed uncertain. In particular, on whom would you rely to distinguish between authentic and fraudulent claims? How will the burden of proof affect such decisions?

9. What scientific assumptions are reflected in Western efforts to retest Chinese acupuncture and other native therapies? Discuss the justification for these assumptions.

SUGGESTED READING AND VIEWING

"Acupuncture." January 1994. *Consumer Reports* 59: 54–59.

Lu, G.-D., and J. Needham. 1980. *Celestial Lancets: A History and Rationale of Acupuncture.* Cambridge, England: Cambridge University Press.

Moyers, B. 1993. "The Mystery of Chi." In *Healing and the Mind.* New York: Doubleday. Transcripted from video. New York: Ambrose Video Publishing.

Pomeranz, B. 1987. "Scientific Basis of Acupuncture." In G. Stux and B. Pomeranz, eds., *Acupuncture: Textbook and Atlas.* Berlin: Springer-Verlag.

Schoen, A. M. (ed.) 1994. *Veterinary Acupuncture: Ancient Art to Modern Medicine.* Goleta, CA: American Veterinary Publications.

CHAPTER 13

Frank Macfarlane Burnet *&* How Animals Make Antibodies

FRED SINGER

☐ *INTRODUCTION*

In the year following World War I, approximately two billion people caught the common flu, and between twenty and forty million people died. In one month, the flu killed 196,000 people in the United States alone. The flu's quick action is legendary. For example, four seemingly healthy women played bridge together one evening during the 1918 pandemic and all went to bed feeling fine, but three never woke up. They were victims of this unheralded rapid-action killer.

Many scientists conducted research on the flu virus in the 1920s and 1930s. During World War II, the search for an effective treatment became more intense because researchers feared another pandemic. Frank Macfarlane Burnet (Figure 13.1) was part of this intense anti-flu campaign. He was motivated by a powerful scientific curiosity to understand how the virus worked and a strong competitive streak, as well as a deeply humanitarian commitment to avert another pandemic at the end of the war. Though he never developed an effective treatment for the flu, Burnet had a tremendous impact on our understanding of the immune system and the nature of disease.

Burnet made two important discoveries regarding immune system function. First, he explained why the immune system doesn't attack an individual's own cells and tissues. Second, he helped develop the clonal selection theory of antibody activity.

Burnet's contributions were closely related to his social and intellectual background. During childhood, he gained a deep appreciation for ecological and evolutionary principles. Burnet used these perspectives in two related ways. First, he understood how disease-causing microorganisms, like other more familiar organisms, experienced their own struggle for existence within their environment—the human body. Second, he was able to make connections between ecological interactions within an ecosystem and immunological interactions within the body. He used Darwin's theory of natural selection as an analogy to interpret how the body can respond, or "adapt," to specific disease organisms.

FIGURE 13.1 Frank Macfarlane Burnet holding the first monkey paralyzed by an Australian polio virus. *Source:* Frank Macfarlane Burnet, *Changing Patterns: An Atypical Autobiography*, 1968.

BURNET'S EARLY YEARS

Frank Macfarlane Burnet was born in 1899 in a small town in Australia. His oldest sister was mentally retarded, perhaps as a result of birth complications, and required an inordinate amount of special care. Burnet's parents attempted to conceal her condition and the children were not allowed to bring friends to the house. His relatively isolated childhood may have led to Burnet's shy personality, which he still retained late in life when he was basking in the glow of scientific achievement.

As a result, Burnet spent a tremendous amount of time by himself, either reading books or wandering around outside, learning about the secrets of nature. He became an avid collector of all natural items, including rocks, butterflies, bird eggs, and freshwater mussels. But his greatest enthusiasm was reserved for collecting beetles, a passion that he shared with another naturalist, Charles Darwin. In his autobiography, Burnet admits that his passion for beetle collection, which peaked while he was in medical school, was probably a replacement for a social life that more outgoing men might enjoy at that age.

As a naturalist and proponent of Darwin's theories, Burnet extended Darwin's great struggle for existence to the microscopic world of disease. He was intellectually prepared to observe infectious agents competing for suitable hosts and responding adaptively to changes in their habitat. He took this unique perspective with him when he entered medical school and maintained it throughout his career.

AN ECOLOGICAL POINT OF VIEW

Burnet applied an ecological point of view to the world of microorganisms. Like Darwin, Burnet was impressed by the tremendous reproductive potential of all organisms, but because he focused his work on microorganisms, Burnet was able to appreciate the incredible reproductive potential of bacteria and viruses in the appropriate environment. In 1940, he argued that the application of the ecological point of view to understanding infectious diseases was the most important attitude change by microbiologists in recent history.

In Burnet's ecological point of view, disease-causing organisms—pathogens— are engaged in the same struggle for existence as other organisms. The difference is that pathogens often make their living *within* a host environment rather than *outside* it. In order to be successful, pathogens still need to deal with the basic necessities of life: food and reproduction. While gathering food and making offspring within their host's body, they must avoid being preyed upon by other organisms or, in the case of many pathogens, the host's immune system.

Burnet described how Australian scale insects invaded California citrus groves. These insects make a living by sucking out the juices from citrus trees, sometimes killing a tree within a year or two. Scale is not a serious problem in Australia, but by 1889, the California lemon crop was threatened with extinction. What was the difference between the two environments?

When the scale insect invaded the new host, there were no natural predators, so the scale insect numbers increased dramatically, threatening to destroy their host. After ecologists successfully introduced a predatory species of ladybird beetle, the numbers of scale insects declined dramatically. Thus a balance was established, with a relatively low number of predators and prey within the habitat.

The ecological and evolutionary perspectives complement each other. Each pathogen is as much the product of adaptive evolution as is the host. Both the pathogen and host share an evolutionary history of living together. The evolution of the immune system is the adaptive response of the host species to generations of pathogens. Burnet argued that when two organisms have evolved together in a host/pathogen relationship, the long-term survival of the pathogen is best served by the development of a pattern of limited infection. Sufficient host material is eaten to keep the pathogen and its offspring alive, but the host is otherwise not seriously injured, allowing the pathogen a greater opportunity to infect new hosts. By killing the host, the pathogen destroys its own environment. This is a very unstable balance because genetic changes in the pathogen may lead to epidemic outbreaks of pathogen-induced disease.

PROBLEM

Burnet argued that pathogens benefit by keeping their hosts alive. Keeping a host alive is costly to the pathogen from an evolutionary perspective, however, because less of the host can be consumed and fewer pathogen offspring are produced. If a genetic mutation arises that allows a pathogen to be transmitted much more easily from one host to another, will the pathogen still benefit as much from keeping its host alive? In the long run, would the offspring of this pathogen be more or less likely to kill their hosts?

EARLY HYPOTHESES FOR ANTIBODY ACTIVITY

One of the most prominent scientists of the late 1800s was Paul Ehrlich, who established a technique for measuring the quantity of antibodies within the blood. Using this technique Ehrlich came to appreciate just how explosively antibodies proliferate following the introduction of an invading molecule or **antigen**. Ehrlich turned his attention to the question of how antibodies form in the body. According to his **side-chain theory** of antibody formation, the surface of a white blood cell bears receptors with a number of different types of side chains to which the antigens bind. Each white blood cell carries the full diversity of side chains that react to different incom-

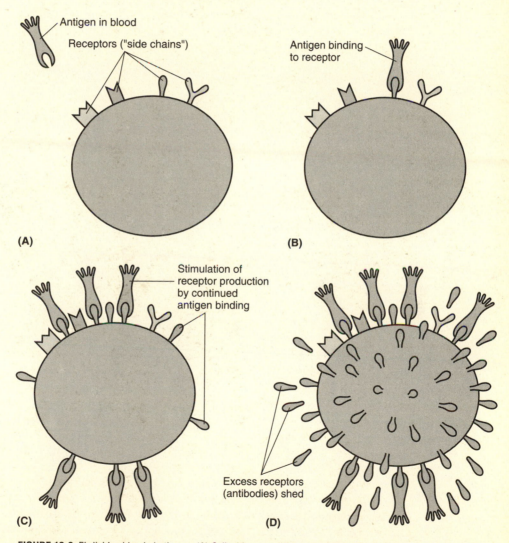

FIGURE 13.2 Ehrlich's side-chain theory. (A) Cell with many different types of side chains. (B) Antigen binds to a receptor, inducing formation of more receptors specific to the antigen. (C) More antigens bind to the receptors, inducing formation of more identical receptors. (D) Receptors are shed into blood as antibodies.

ing antigens. When the antigen is linked to the side chain, the white blood cell produces multiple copies of the correct receptor, which it dumps into the bloodstream as antibodies (Figure 13.2).

Organic chemists cast serious doubt on Ehrlich's theory based on the physical and molecular properties of these newly isolated antibodies. They found that animals could generate an almost unlimited number of antibodies, including antibodies that were specific to new, synthetic molecules. How could animals produce unique antibodies with side chains specific to molecules that had never existed on earth before? As the number of different types of known antibodies increased, it became clear that there was simply not enough space to fit all these different types of side chains on the surface of each white blood cell.

In response, a group of organic chemists proposed an alternative theory—the **template theory**. According to this theory, the antigen is carried in the body to the

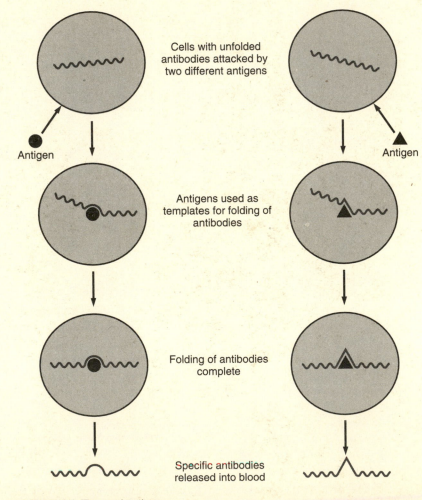

FIGURE 13.3 The template theory.

site of protein formation, where it serves as a template upon which the antibody molecule is constructed. The antibody molecule is synthesized upon the surface of the antigen so that the molecular structure of the antibody is complementary to the antigen. The antibody, when released into the blood, would be a perfect fit to the antigen. Hence there was no need for an innate diversity of antibodies—instead, diverse antibodies were produced by the body in response to its experiencing new antigens (Figure 13.3).

Some biologists objected to the template theory. One reason was a distrust by biologists of the purely molecular approaches used by the chemists. Although they recognized the importance of chemistry, immunologists criticized the template theory because it ignored important biological phenomena associated with antibody activity.

Problems with the Template Theory

During much of his career, Burnet supported the template theory, but he realized that it had potentially damaging problems. One problem was that animals failed to produce antibodies to particular antigens under some specific conditions. For example, under normal circumstances an animal doesn't make antibodies to itself. The immune system can distinguish, in Burnet's words, between self and not-self, resulting in what we call **immunological tolerance**. The template theory had difficulty explaining how the immune system was able to make this discrimination.

Two experiments demonstrated that in very early stages of development, an animal may not yet have immunological tolerance. In one experiment, Burnet was unable to evoke an immunological response in chick embryos, despite trying to do so with three common types of antigens that caused powerful immunological responses in slightly more mature chickens. Thus there was a stage in development when chicks made no immune response to foreign antigens. In a second experiment, a colleague of Burnet's demonstrated that fraternal (nonidentical) twin calves exchanged blood through a common placenta, maintaining a mixture of two different types of blood antigens in the uterus without reacting to the antigens provided by the other calf's blood. Again, this provided evidence that, early in development, animals tolerated foreign antigens. Burnet's early work on immunological tolerance ultimately paved the way for developing successful techniques for organ transplants. Accordingly, he was awarded the Nobel Prize in 1960, along with the British zoologist, Peter Medawar.

During World War II, Medawar studied how skin grafts were accepted or rejected by burn victims of incendiary bombs. After the war, he used Burnet's findings as the basis for a series of experiments on mouse cells. Instead of studying antibody production, he used acceptance or rejection of a skin graft to measure immune response. When he grafted skin from three-week-old mice of one strain onto the skin of mice from a second strain, the graft was always rejected. But if he injected spleen cells from a mouse of the first strain into a newborn mouse of the second strain, then attempted the same skin graft after the mouse was three weeks old, the graft was accepted. As Burnet had predicted, the mice developed immunological tolerance of the graft as a result of exposure to the cells very early in development. The host treated the skin graft as "self" rather than foreign tissue.

Medawar's findings cast doubt on the template theory. According to the template theory, initial exposure to antigens stimulated production of specific antibodies. If this theory were correct, mice exposed to a foreign tissue early in life should reject a skin graft of the same foreign tissue three weeks later.

Burnet and his colleagues raised other equally important objections to the template theory. First, the effectiveness of antibodies seemed to improve over the course of the immune response. Second, it was becoming clear that antibody production continued after the antigen was no longer present in the circulation, leading to an unexplainable situation where nonexistent templates were producing antibodies. A final problem for the template theory was that the secondary response to the same antigen was characteristically much more rapid and intense than the initial (primary) response—that is, the immune system showed memory (Figure 13.4). Why should reintroducing the antigen (template) give rise to more antibody molecules than did the initial introduction of the template?

PROBLEM
Four problems for the template theory are (1) immunological tolerance, (2) increased effectiveness of antibodies, (3) continued response after antigens are no longer present, and (4) immunological memory. Revise the template theory in a way that could account for each finding.

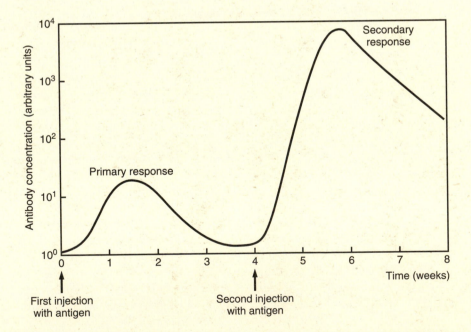

FIGURE 13.4 Differences in timing and magnitude of antibody formation in response to primary and secondary exposure to a given antigen.

Despite its problems, most scientists in the 1950s supported the template theory because it explained why antibodies are specific for a tremendous diversity of antigens. Burnet's objections did, however, make many scientists aware of difficulties with the template theory and paved the way for considering new solutions to these problems.

DEVELOPMENT OF A NEW THEORY

In 1955, Niels Jerne resurrected Ehrlich's side-chain theory in a slightly novel form, which he called the **natural selection theory** of antibody formation. He argued that diverse antibodies are normally formed by the animal in small quantities. When an antigen of a particular type enters the blood, it eventually encounters an antibody of appropriate specificity and is bound to it. This antigen/antibody complex then moves to specialized cells that reproduce the specific antibody in large numbers. These cells have the ability to faithfully reproduce whatever antibody is brought to them, though occasional mistakes are made. These mistakes may result in an antibody that binds even better to the antigen (is a better fit). Jerne had no definite opinion about when the initial population of antibodies was produced. But he did deal with the problem of immunological tolerance by suggesting that newly created antibodies that attach themselves to tissues in the animal's body are removed from circulation, and thus not available for reproduction.

> ### PROBLEM
> Reconsider the four problems raised by Burnet and his colleagues. Which does the natural selection theory successfully deal with? Which are still problematic?

Upon reading Jerne's article, Burnet had a mixed reaction. Personally, Burnet was annoyed with Jerne, who had recently attacked Burnet's book on immunology as being overly philosophical. From a biological perspective, Burnet was also concerned that Jerne's theory could not explain the origin of antibody diversity or how antibodies were replicated so accurately by the cells. He did, however, appreciate that Jerne's selection theory dealt with many of the template theory's problems. Burnet was also enthusiastic about the Darwinian analogy that ran through Jerne's theory, and, as a beetle collector with thousands of different species in his collection, he could easily imagine that a body could produce the innate diversity of natural antibodies required for Jerne's theory to work. This appreciation for diversity in biological systems distinguished him from most medical researchers.

While Burnet was thinking about Jerne's paper, researchers in his institute were beginning to produce evidence that white blood cells carried some immunological properties. These findings, linked with Jerne's paper, helped Burnet formulate the concept of **clonal selection**. He proposed that each cell, based on its genetic composition, produces characteristic receptors on its surface that are complementary to an antigen. If the receptors bind a foreign antigen, the cell is induced to proliferate. Each immunologically active cell (lymphocyte) is genetically constrained to produce

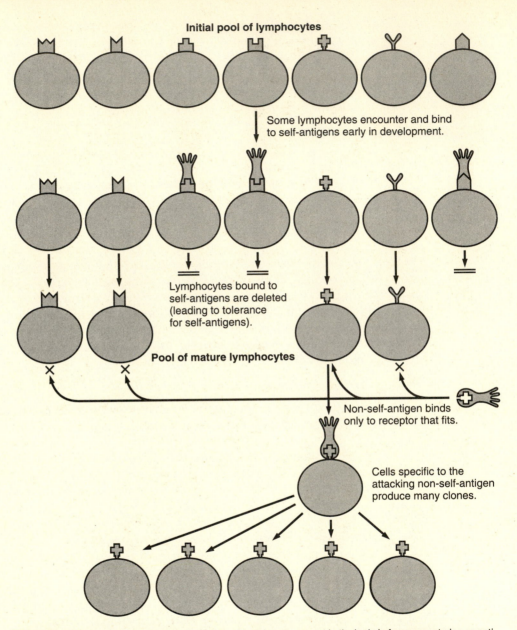

Initial pool of lymphocytes

Some lymphocytes encounter and bind to self-antigens early in development.

Lymphocytes bound to self-antigens are deleted (leading to tolerance for self-antigens).

Pool of mature lymphocytes

Non-self-antigen binds only to receptor that fits.

Cells specific to the attacking non-self-antigen produce many clones.

FIGURE 13.5 Clonal selection theory. Each class of lymphocyte present in the body before encountering an antigen has a distinct receptor for a specific antigen. Lymphocytes with receptors that bind to self-antigens are eliminated early in development, assuring tolerance of self-tissue. A mature lymphocyte, upon binding an antigen, is stimulated to go through a series of mitotic divisions, yielding a clone of identical progeny, all with receptors specific to that particular type of antigen. Some of these progeny develop into effector cells that eliminate the antigens, while others mature into memory cells, which remain in circulation, ready to respond to the next challenge by the same type of antigen.

only one kind of receptor on its surface. After successful reproduction, the receptors on the surface are then shed into the blood as antibodies (Figure 13.5).

Burnet described two advantages to his revision of Jerne's theory. First, it seemed to deal with all the problems of the template theory and with his objections to Jerne's natural selection theory. Immunological tolerance arises because entire clones of lymphocytes are simply deleted very early in development if they match the tissue of the individual (Figure 13.5). The binding abilities of antibodies improve over time because the antigens act as agents of selection, "choosing" cells with receptors that are literally the best fit, and inducing them to proliferate. Once the lymphocytes have gone through a series of reproductive cycles, they will make antibodies even after the antigen is gone by continuing mitosis and shedding their receptors. Finally, the secondary response is more rapid and powerful than the primary response because there are more antigen-specific cells in the blood as a result of the initial antigenic attack.

The second attraction of this theory for Burnet was more philosophical, reflecting his admiration for Charles Darwin. His theory of immune response was analogous to Darwin's theory of natural selection. Furthermore, as a microbiologist, he presented a theory demonstrating that the microscopic world behaved in much the same way as the macroscopic world more familiar to most naturalists. Burnet saw the immune response as a Darwinian world in miniature, with the lymphocytes forming a population within a community, undergoing differential survival and reproduction in relation to their individual fitness. In this case fitness is determined by how well the receptors on the surface of the lymphocytes "fit" to the antigen that enters the environment. Thus fitness, as in Darwin's world view, reflects changes in the environment. In addition, accidental changes (mutations) in the genetic makeup of the cells introduce novel variation into the population, serving as raw material for selection.

Despite the tremendous success Burnet was enjoying, he still retained some of his early childhood insecurities. He had already proposed two incorrect hypotheses of antibody function, both of which involved template types of mechanisms. Thus he published his clonal selection theory in an obscure Australian journal so as not to embarrass himself too badly if he was wrong once more. Additionally, he had a great deal of patriotism for his homeland and wanted this idea to see the first light of day in Australia.

Burnet and Jerne became admirers of each other's work. Burnet considered Jerne the most intelligent living immunologist, and when Jerne received a Nobel Prize in 1984, he sent Jerne a congratulatory letter stating that their joint theory was more deserving of a Nobel Prize than was the tolerance research for which he and Medawar had shared the prize. Jerne, in addressing a symposium on antibodies, congratulated Burnet for stimulating a great proliferation of immunologists and, later, for hitting the nail on the head with his clonal selection theory.

☐ *EPILOGUE*

Application of Burnet's ecological point of view is a departure point for problems that are challenging medical researchers today. For example, smallpox is a horrible disease that causes weeping sores, high fever, and, often, death. As a result of extensive vaccination programs, the last case of smallpox was reported in 1978. Why can medical researchers develop effective preventions or treatments for some diseases yet be frustrated in their efforts against others, such as influenza or AIDS? Why are some viral diseases more difficult to control than others?

Some viruses are particularly hardy and can withstand long periods of time away from the host. Others are effective travelers, moving from host to host with relative ease. Finally, viruses vary in the speed with which they evolve, with some types changing their genetic identity very frequently.

The smallpox virus is one of the largest and most resilient viruses known, as it can live apart from its host for decades. A medical researcher recently published a paper in which he recommended that archaeologists who worked on mummies should be vaccinated against smallpox, given the virus's ability to survive in cool, dry conditions for many years. It is also easily transmitted from one host to the next. The virus is genetically very stable, however, which made it susceptible to eradication by the major worldwide vaccination effort.

In contrast, the influenza virus, which so interested Burnet, has a mutation rate about 100 times greater than many other viruses. This high mutation rate exists because the genetic material of the flu virus is RNA. In contrast to DNA viruses like smallpox that have effective repair systems, RNA viruses cannot correct copying errors during the process of replication. Each flu outbreak is caused by a virus with slight changes in its surface antigen, rendering it unrecognizable to previously formed memory cells and insensitive to immunity produced by vaccines made the previous year.

Periodic influenza pandemics arise from two major causes. First, the influenza virus has a segmented genome, with each of the eight loosely bound segments responsible for producing one or two viral proteins. The loose connection between segments allows them to come apart and rearrange with segments from other nearby viruses. If they recombine with segments from viruses in other animals, there can be significant changes to the surface antigens (antigenic shift) that result in the new virus being completely unrecognized by the immune system of the human host it encounters. Second, humans provide an environment in which this genetic reassortment is very likely to occur. Chickens, pigs, and ducks are excellent vectors for the virus, as they harbor the virus within their guts but don't get sick from it. Particularly in China, chickens, ducks, and pigs are commonly raised together; all are likely to get infected with human flu virus, and if they are concurrently harboring the virus from their own species, this form of reassortment may occur. With the recent increase in fish farming in China, which involves feeding hen feces to pigs and fertilizing the fish ponds (which are also duck ponds) with pig manure, there is cause for serious concern that the frequency of gene reassortment will increase over the next few years. Given that ducks are excellent long-distance migrants, they may spread the virus throughout the world very quickly.

As a final example, the HIV virus, which causes AIDS, is relatively fragile outside the body, losing its infectious properties within two hours of exposure to the air. It is also very difficult to transmit, requiring the bodily fluids of a host individual to enter into another person's tissues. HIV, however, has two features primarily responsible for its deadly effect on humans. First, it infects cells of the immune system, seriously restricting the immune system's ability to mount any serious defense. Second, once the viral RNA enters the host cell, it manufactures a double strand of viral DNA that inserts itself into the nucleus of the host cell. There it remains, often for many years, until activated by factors at this point still unknown to immunologists.

There are several implications to this unusual mode of reproduction. Once the HIV genetic material is integrated into the host DNA, the immune system does not appear to recognize it as a foreign antigen. The long latency period between initial infection and symptoms of serious illness increases the chance that the infected person will transmit the disease to one or more hosts. Additionally, being an RNA virus, HIV has a very high mutation rate; in fact, an individual may have more than one kind of HIV variant as a result of mutation events that occurred after the initial infection.

Burnet's evolutionary perspective is now being adopted by medical researchers to combat a new and potentially very deadly problem—that of the evolution of antibiotic resistance. Fifty years ago, Burnet warned the medical community that antibacterial drugs should be used with great caution, as their unsupervised use was creating a novel environment selecting for new strains of pathogens. Today, the medical community is finding that many of its most potent antibacterial drugs are no longer effective against the new strains of antibiotic-resistant bacteria that have adapted to their novel environment.

QUESTIONS AND ACTIVITIES

1. What does this case show about the following aspects of doing biology?
   — the use of analogies
   — importance of perspective and background
   — usefulness of a natural history background
   — consideration of alternative scientific theories
   — communication within the scientific community

2. Burnet uses the example of ladybird beetles, scale insects, and lemon trees as an analogy to how the immune system functions. Which characters within the immune system are analogous to ladybird beetles, scale insects, and lemon trees? How do they interact with each other? How does this analogy help you understand the immune system?

3. Burnet freely credited Jerne with stimulating the clonal selection theory. In contrast, Jerne did not even cite Ehrlich in his original paper. How would Jerne have known about Ehrlich's work? Propose some reasons that could account for why Jerne did not cite Ehrlich. What does this case reveal about how theories build upon previous theories?

4. The central dogma of molecular biology states that DNA makes RNA, which makes protein. Given that antibodies are protein (globulin) molecules, is Jerne's natural selection theory compatible with the central dogma? Why or why not?

5. Regarding natural selection of organisms and clonal selection of lymphocytes, answer the following:
   a. What is being selected?
   b. What is meant by "fit"?

   c. What is the role of mutation?

   d. What is the role of diversity?

   In what ways does this analogy help you understand how the immune system functions?

6. Propose a set of practical guidelines for dealing with the increase in antibiotic resistance in disease organisms. In your answer, consider the following information.

   a. Some types of antibiotics are broadscale, affecting many types of bacteria, while others are very specific.

   b. Some ranchers routinely put antibiotics in cattle feed.

   c. Today, many antibiotics must be given in much larger doses than previously in order to be effective.

   d. Recently, several people died in hospitals from antibiotic-resistant infections that they picked up in the hospital.

SUGGESTED READING

Ada, G. L., and S. G. Nossal. 1987. "The Clonal-Selection Theory." *Scientific American* 257: 62–69.

Burnet, F. M. 1953. *Natural History of Infectious Disease*. (2nd ed.) Cambridge, England: Cambridge University Press.

Burnet, F. M. 1957. "A Modification of Jerne's Theory of Antibody Production Using the Concept of Clonal Selection." *Australian Journal of Science* 20: 67–69.

Burnet, F. M. 1968. *Changing Patterns: An Atypical Autobiography*. Melbourne, Australia: William Heinemann.

Henig, R. M. 1994. *A Dancing Matrix*. New York: Random House.

Jerne, N. 1955. "The Natural-Selection Theory of Antibody Formation." *Proceedings of the National Academy of Science* 41: 849–857.

Sexton, C. 1991. *The Seeds of Time: The Life of Sir Macfarlane Burnet*. Melbourne, Australia: Oxford University Press.

Silverstein, A. M. 1989. *A History of Immunology*. San Diego, CA: Academic Press.

CHAPTER 14

Niko Tinbergen & the Mating Behavior of Sticklebacks

JOEL B. HAGEN AND FRED SINGER

☐ INTRODUCTION

If you visited a lake as a child, you may remember being fascinated by the bluegill and pumpkinseed sunfish swimming under the dock. Sometimes floating almost motionless in the shade, other times darting away from some unseen disturbance, their behavior probably appeared completely random. Careful study, however, reveals that each of these closely related species has a distinctive pattern of behavior, particularly during the breeding season. Male pumpkinseeds are solitary animals. Equipped with hard, bony mouthparts well adapted for nipping potential predators, each male is able to protect his own nest site from marauding catfish who might eat the eggs. In contrast, the small, delicate mouthparts of bluegills pose little threat to nest-raiding predators. For protection, bluegills join together in breeding colonies of 50 to 100 males. Strength in numbers may be a good strategy against nest-raiding catfish, but being social poses other problems for male bluegills. Each male vigorously defends a small territory surrounding his nest from the other males, who might slip in and fertilize the female's eggs first.

The threat of having another male fertilize the eggs in his nest is a serious problem for territorial males. Some male bluegills who don't build nests specialize in sneaking into other males' nests to spawn. Perhaps not surprisingly, these interlopers are often small and furtive. Another type of male mimics female coloration and behavior to prevent detection by nest-building males.

How have these alternative forms of behavior evolved and how are they adaptive? Are complex forms of social behavior learned or do they have some genetic basis? Are there principles of animal behavior that can be generally applied, perhaps even to humans? Such questions intrigued the Dutch biologist Niko Tinbergen, and he devoted his career to answering them. Together with Konrad Lorenz, he developed the new field of *ethology* to study the biological basis of behavior. For their pioneering studies of animal behavior, Tinbergen, Lorenz, and another ethologist, Karl von Frisch, were awarded a Nobel Prize in 1973.

FIGURE 14.1 Niko Tinbergen. *Source:* M. S. Dawkins, T. R. Halliday, and R. Dawkins, eds., *The Tinbergen Legacy*, 1991.

THE MAKING OF AN ETHOLOGIST

Niko Tinbergen (Figure 14.1) was born into a family where learning was encouraged. The family had the unique distinction of having two children win Nobel Prizes. Older brother Jan won the prize for economics in 1969, four years before Niko was awarded his prize.

Despite his intellectual abilities, Tinbergen was a somewhat indifferent student. He excelled in courses that interested him but ignored those he found boring. Two other activities competed for his attention. Tinbergen was an avid athlete, competing internationally in field hockey. Throughout his career, he remained physically active, and he always enjoyed working outdoors. Like many animal behaviorists, he was also drawn to natural history at an early age. His family encouraged this extracurricular interest, and as a boy he spent much of his time on field trips collecting animals.

His interest in natural history took a serious turn when he entered the University of Leiden. His teachers quickly recognized his abilities, and Tinbergen joined a small, but active, group of Dutch animal behaviorists. After completing his Ph.D., he taught zoology at the university. Although still in his twenties, Tinbergen was already building an international reputation as a talented scientist.

During the early 1930s Tinbergen met the Austrian biologist Konrad Lorenz. At the time, Lorenz was trying to establish a new approach to animal behavior, which he called **ethology**. Tinbergen's observational and experimental skills complemented Lorenz's more theoretical approach. The two men forged a lifelong friendship, and together they made ethology an important field of biological study.

This activity was temporarily interrupted during World War II, when Tinbergen was imprisoned for his resistance to the Nazis. After the war, Tinbergen left Holland to take a teaching position at Oxford University. This was an important move for Tinbergen, because Oxford had become a center of field biology. There he joined an outstanding group of biologists who were interested in animal behavior, ecology, and population genetics (see Chapter 1).

SEX AND STICKLEBACKS

Choosing the right organism to study is an important ingredient in research. Although Tinbergen studied many species, one of his favorite subjects was the three-spined stickleback (*Gasterosteus aculeatus*), a small fish named for its protective dorsal spines. As a boy, Tinbergen had often caught sticklebacks in ditches near his home, and, he later recalled, "I soon discovered that in choosing these former pets I had struck oil." Sticklebacks are hardy animals, easy to collect and keep in the laboratory. Unlike mammals whose behavioral characteristics are complicated by learning, this fish always seemed to respond in a predictable way to stimuli. According to Tinbergen, sticklebacks provided excellent examples of what he referred to as "automatic" or "purely instinctive" behavior. For all of these reasons, they were a good choice for behavioral experiments. Perhaps with tongue in cheek, Tinbergen claimed that "to us this little fish is what the rat is to many American psychologists."

Tinbergen was particularly interested in the mating behavior of sticklebacks. During most of the year sticklebacks live together in schools. At the beginning of the breeding season, however, males become territorial. On the bottom of the stream, each male builds a nest out of algae and plant material, and he actively drives all

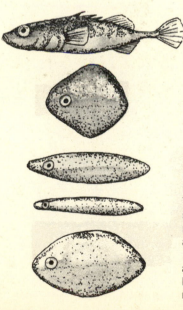

FIGURE 14.2 Models used in experiments on fighting behavior in territorial male sticklebacks. Territorial males attacked models with red undersides, even if they did not look like fish. Realistic models without the red coloration were rarely attacked by territorial males. *Source:* from *Biology: the Network of Life*, by Michael C. Mix, Paul Farber, and Keith I. King. Copyright ©1992 by Michael C. Mix, Paul Farber and Keith I. King. Reprinted by permission of HarperCollins Publishers.

other males and nonreceptive females away from the surrounding area. Physiological changes accompany this change from nonaggressive schooling to aggressive territoriality. The abdomens of territorial males turn bright red. In contrast, egg-bearing females turn glossy silver, and their abdomens become swollen with eggs.

Through a series of elegantly simple experiments, Tinbergen demonstrated that the color red serves as a stimulus for aggressive behavior. Almost any red object placed in front of a territorial male elicited an aggressive display. Wooden models, which bore little resemblance to male sticklebacks, provided this stimulus as long as they had red undersides (Figure 14.2). In fact, Tinbergen noticed that males often reacted aggressively when red mail trucks passed by the window of his laboratory. In contrast, perfectly shaped models of male fish that lacked red coloration were rarely attacked.

PROBLEM

Consider the models shown in Figure 14.2. What characteristics of male fish are represented by these models? What male characteristics might be missing in a wooden model? How could Tinbergen modify his models to test whether the missing characteristics also stimulated territorial behavior?

According to Tinbergen, the aggressive, territorial behavior of male sticklebacks illustrated two important characteristics common to all innate behavior. First, the behavior had a physiological basis. Males acted aggressively only when their

FIGURE 14.3 A male and female stickleback performing the sequence of courtship behavior that leads to mating. Males often court and mate with several different females. *Source:* from *Biology: the Network of Life*, by Michael C. Mix, Paul Farber, and Keith I. King. Reprinted by permission of HarperCollins Publishers.

abdomens turned red and they were physiologically ready to breed. Second, the behavior was always a response to a specific stimulus. A single male trait, red coloration, stimulated attacks. All other male characteristics seemed to be irrelevant.

Even some very complex forms of behavior could be thought of as chain reactions of simple stimulus-response mechanisms. The zigzag courtship dance between a male and a receptive female illustrated this more complex mechanical pattern (Figure 14.3). When a female enters a breeding territory, the male swims vigorously toward her. Such an aggressive approach causes most nonreceptive females to flee, but an egg-bearing female responds by facing the approaching male and exposing her swollen abdomen. This simple response acts as a new stimulus that switches the male from aggression to courtship. He responds by leading the female in a zigzag pattern toward the nest.

Once the pair arrive at the nest, the male turns on his side and points to the entrance with his snout. The female enters the nest and remains there with her head and tail protruding from either end. Trembling, the male prods the female's tail several times, which causes her to release her eggs. After spawning, the female swims out of the nest and the male enters to fertilize the eggs.

After mating is complete, the male again becomes aggressive and chases the female out of the territory. He then begins to court other egg-bearing females that enter his nest site. After fertilizing several clutches of eggs, the male stops courting and begins to show parenting behavior. The eggs seem to provide a stimulus for fanning, which the male does to provide them with oxygen. After the eggs hatch, the male protects the young. Females take no part in this parenting behavior.

PROBLEM
Propose a testable hypothesis about what characteristic of the eggs stimulates the switch from courtship to parenting behavior (fanning eggs) in male sticklebacks. How would you test this hypothesis?

Tinbergen's careful description of the courtship dance highlighted several important features of mating behavior. The dance was complicated, but it comprised a series of quite simple steps. Each step involved a specific stimulus from one fish followed by an equally specific response from the other. Finally, the type of response depended not only upon the stimulus but also upon the physiological state of the fish. For example, the color red was an important stimulus, but it caused different responses in males and females. Males would attack red-colored models, even those that did not look like fish. Gravid females would follow these same unrealistic red models through the zigzag courtship dance, even if a nest was not present.

THE ETHOLOGICAL METHOD

Unraveling the details of the zigzag dance required several different scientific methods. Careful observation is the starting point for all behavioral research, and it was particularly important for ethologists. Tinbergen cautioned his students not to begin experiments until they thoroughly understood the natural behavior of the animals they studied. This meant spending hours watching behavior without disturbing the animals in any way. Often this was done in the field, but in the case of the sticklebacks, Tinbergen was able to observe the fish in aquaria.

After becoming thoroughly familiar with a complex behavior, the ethologist could begin to analyze its component parts. Often this involved creating an **ethogram**, which was a detailed list or description of the types of behavior performed by a species. Ethograms generated questions about the frequency, intensity, sequence, and duration of simple behavioral characteristics. At this point in the study, the ethologist might design experiments to test hypotheses. Although experiments played a critical role in discovering the causes of behavior, Tinbergen stressed that experimentation must always be balanced with careful observation.

> ### PROBLEM
> Reconsider the simple behavior of fanning eggs that males perform after mating. This behavior, which provides the eggs with oxygen, is crucial for the survival and development of embryos. Propose a testable hypothesis about how eggs stimulate the frequency, intensity, or duration of this behavior. Design a simple experiment to test your hypothesis.

The final step in the ethological method was comparing the same types of behavior among closely related species. By comparing similarities and differences among species, ethologists could sometimes draw inferences about how the behavioral characteristics evolved. Behavioral characteristics could also sometimes be used as important pieces of evidence for determining evolutionary relationships among related species.

NATURE VERSUS NURTURE

Ethologists believed that all behavior had a genetic component that is passed from parent to offspring. Some simple behavioral characteristics seemed so rigid and uniform that they appeared to be inherited much like fins, gills, or spines. Just as all of the sticklebacks that Tinbergen studied had three dorsal spines, all males and females seemed to perform the courtship dance in the same way. In both cases, the characteristics were adaptive, and natural selection tended to eliminate unusual variants. Just as a mutant stickleback with missing spines would be quite defenseless against predators, a stickleback that could not correctly perform the courtship dance would be unsuccessful at mating.

Tinbergen thought that courtship rituals had evolved by natural selection. Consequently individuals almost always mate with members of their own species and avoid mating with members of other species. For example, ten-spined sticklebacks (*Pungitius pungitius*) performed a courtship dance very similar to that of three-spined sticklebacks. Rather than having red abdomens, however, male ten-spined sticklebacks had black undersides. Several of the individual dance steps were also slightly different in this species. Thus both anatomical and behavioral characteristics were adaptations to prevent interbreeding. In order for this reproductive isolation to work, all members of each species would have to perform the appropriate courtship ritual. A female three-spined stickleback would produce no offspring if she mistakenly responded to the black abdomen and courtship behavior of a male ten-spined stickleback.

In many simpler animals, almost all behavioral characteristics seemed to be instinctive. Tinbergen often spoke of sticklebacks as if they were machines that automatically reacted to stimuli. Using a modern figure of speech, ethologists believed that each species was "hardwired" to perform characteristic innate forms of behavior. Even in humans, which Tinbergen also studied, one could find some simple, instinctive behavioral characteristics. For example, breast-sucking seems to be an instinct that is automatically done by all newborn infants. Perhaps even complex forms of learned behavior had evolved from such instincts.

This emphasis on instinct put ethologists at odds with many American psychologists, who were primarily interested in learning. Using rewards and punishments, experimental psychologists could drastically modify behavior. Rats and pigeons could be trained to push buttons or pull levers to receive food or avoid electrical shocks. Impressed with these results, some psychologists believed that all behavior was learned in this way. They proposed that at birth the animal's brain was like a blank slate. Through experience, both painful (punishments) and pleasant (rewards), the animal learned to behave in appropriate ways.

This "nature versus nurture" controversy polarized animal behaviorists during the 1950s. Ethologists criticized psychologists for using highly artificial experiments and for studying only a few species of animals: rats, pigeons, chickens, and primates. What could be learned about natural behavior from timing rats as they ran through mazes? Psychologists criticized ethologists for ignoring the development of behavior as young animals grew into adults. How could ethologists be so sure that learning did not occur, even in many simple animals?

PROBLEM

Behavioral psychologists stressed two forms of learning that might be used even by simple animals: copying and trial and error. Design an experiment to test whether copying is involved in the development of stickleback courtship behavior. Design another experiment to test whether trial-and-error learning plays a role in the development of stickleback courtship behavior.

By the 1960s, the nature versus nurture controversy had largely subsided. Both ethologists and psychologists realized that a sharp distinction between instinct and learning was misleading. Almost all forms of behavior involved both elements. Tinbergen himself pointed out that four different questions could be asked about behavior. How was it caused by nerves and hormones? How did it develop in the growing organism? How was it adaptive in a particular environment? And how had it evolved? These questions were equally important, but a single scientist might not be able to answer all of them. Studying each question required a different set of methods and, perhaps, a slightly different perspective on what was important about animal behavior.

Like all great scientific achievements, Tinbergen's early studies of sticklebacks raised interesting new questions as they answered old ones. By posing these questions, his books and articles continue to stimulate creative research. A new group of behavioral ecologists, many of them Tinbergen's students, has criticized, expanded, and refined his explanations of the adaptive function of courtship behavior and how it evolved in different environments. This ongoing research tradition is perhaps Tinbergen's greatest legacy.

☐ EPILOGUE

Unlike Tinbergen, who emphasized the ways in which courtship behavior was the same in all male or female sticklebacks, behavioral ecologists are now more interested in behavioral differences among individuals of each sex. Thus the focus of study has shifted from the level of the species to the level of individuals within a population or species. What caused this change from traditional ethology to behavioral ecology?

Beginning in the mid-1960s, evolutionary biologists began to emphasize that natural selection usually operates at the level of individuals rather than whole populations or species. If the individual is the unit of selection, it makes sense to ask whether behavioral differences among individuals are due to genetic differences. If there is a genetic basis for behavioral differences, we would expect that natural selection would favor some behavioral traits over others. Like physical characteristics, behavior should contribute to individual fitness.

This emphasis on natural selection and individual fitness was accompanied by new research methods. Behavioral ecologists, including Tinbergen later in his career, studied large numbers of animals in natural populations. In many cases they marked or tagged individuals. This method allowed the scientists to study variation in behavior among individuals and quantify its consequences for survival and reproductive success. This was quite different from Tinbergen's early study of sticklebacks, which involved observing only a small number of fish living in aquaria.

Tinbergen generally adopted a cooperative world view in which males and females harmoniously courted, mated, and raised offspring. The primary focus of this complex process was the male's behavior. Females were portrayed as passively following the male's lead during the zigzag dance. Although he recognized conflicting drives (as evidenced by the aggressive components of the zigzag dance), Tinbergen emphasized the coordination of male and female behaviors, leading to successful reproduction. Courting and mating always seemed to take place in a predictable way.

Behavioral ecologists view courtship behavior quite differently. By focusing on how different individuals may benefit from different types of behavior, biologists now realize that there is often a strong element of competition between males and females. When male and female interests don't converge, the two sexes may not always work harmoniously. In contrast to Tinbergen's view of females as followers, behavioral ecologists now recognize that female sticklebacks play an active role in determining the reproductive success of both sexes.

Females actively choose males on the basis of where their nests are located. Males who build nests in well-protected areas that are relatively free of predators tend to be most successful in attracting mates. Because the quality of territories is so important, males must compete for areas camouflaged by protective layers of filamentous green algae. The red coloration of territorial males that Tinbergen emphasized may be correlated with a male's competitive abilities, but it is the quality of the territory that first attracts the female.

A more dramatic female behavior that Tinbergen did not discuss is egg cannibalism. Females often attack males' nests and eat eggs or developing young. This benefits the female because she gets a nutritious meal and eliminates the eggs of her rivals. It appears that cannibalistic females are very successful in mating with males whose nests they have previously raided. Egg cannibalism, however, is always detrimental to the male's reproductive success. Not surprisingly, males have evolved several types of behavior to deter egg cannibalism. Males frequently attack marauding females with vigorous bites and threatening displays. A male may also distract cannibalistic females by feeding outside his territory. Finally, because egg cannibals often travel in roving gangs, males often court single females while avoiding females in groups.

Egg cannibalism is a striking example of how stickleback behavior may depart from Tinbergen's simple description of male-female cooperation. Other recent studies call into question Tinbergen's description of the zigzag dance as a "species-specific" behavior pattern. When behavioral ecologists studied five genetically distinct populations of three-spined sticklebacks, they discovered important differences in the courtship ritual. In two populations, females rather than males initiated courtship, and the zigzag component was either missing or very inconspicuous. In both of these populations, groups of up to 300 females and immature males attacked the nests and cannibalized the young. Presumably, a history of cannibalism by females in these two populations has favored males who do not advertise their nest sites with conspicuous courtship displays.

In three other populations where the males did initiate courtship with the conspicuous zigzag dance described by Tinbergen, cannibalism was never observed. In this social environment, males can use conspicuous courtship displays without the risk of losing their fertilized eggs. Thus courtship behavior appears to be an adaptive characteristic that may evolve differently in different social environments. By comparing different populations within a species, behavioral ecologists have successfully tested hypotheses about the adaptive significance of behavior.

QUESTIONS AND ACTIVITIES

1. What does this case show about the following aspects of doing biology?
   — experimental and nonexperimental methods
   — disciplinary boundaries and interdisciplinary problems
   — how different theoretical perspectives may influence the interpretation of data
   — how research problems and methods change
2. Tinbergen was a very careful observer, so he probably saw individual variation in the way different sticklebacks performed the courtship dance. Why do you think he did not emphasize this variation when he described the behavior?

3. In his descriptions of courtship behavior, Tinbergen emphasized cooperation between the male and female to ensure successful reproduction. Behavioral ecologists often emphasize competition between the sexes. How have both of these perspectives contributed to understanding courtship behavior?

4. Unlike his laboratory studies of sticklebacks, Tinbergen's equally famous research on seagulls and other birds was done on wild populations. These studies involved carefully observing the behavior of many individual birds in their natural habitats. What might be the advantages and disadvantages of studying animal behavior in the field versus the laboratory?

5. Tinbergen and other ethologists used terms like "courtship" and "dance" to describe certain aspects of mating behavior in sticklebacks. More recently, some behavioral ecologists have used the term "cuckoldry" to describe the situation where the eggs in one male's nest are fertilized by another male. The male who fertilizes the eggs is sometimes referred to as a "sneaker." How appropriate are these analogies to human behavior? What are some possible benefits or dangers in using such analogies?

6. Although Tinbergen and other ethologists usually studied nonhuman animals, they believed that their research might provide some insights into human behavior. How might studying bluegill sunfish or sticklebacks help us to understand human behavior?

SUGGESTED READING

Baerends, G., C. Beer, and A. Manning. (eds.) 1975. *Function and Evolution in Behaviour: Essays in Honour of Professor Niko Tinbergen, F.R.S.* Oxford, England: Clarendon Press.

Burkhardt, R. W. 1981. "On the Emergence of Ethology as a Scientific Discipline." *Conspectus of History* 1: 62–81.

Dewsbury, D. (ed.) 1989. *Leaders in the Study of Animal Behavior: Autobiographies of the Founders.* Chicago: University of Chicago Press.

FitzGerald, G. J. 1993. "The Reproductive Behavior of the Stickleback." *Scientific American* 268(4): 80–85.

Tinbergen, N. 1951. *The Study of Instinct.* Oxford, England: Clarendon Press.

Tinbergen, N. 1952. "The Curious Behavior of the Stickleback." *Scientific American* 187(6): 22–26.

CHAPTER 15

J. B. S. Haldane *&* the Evolution of the Hardy-Weinberg Model

FRED SINGER

☐ *INTRODUCTION*

Many of you have enjoyed taking a long automobile trip. Perhaps your destination was a large city, but in other cases, large cities were obstacles that you needed to navigate as quickly and painlessly as possible. In either case, you may have pulled out a piece of paper with gray, blue, and red lines drawn on it and engaged in the time-honored sport of trying to drive a car and read a map at the same time.

A map is a type of **model**. Like all models, it allows the user to make hypotheses about reality. By using the model, you should be able to arrive at or near your destination. Or, if your goal was to bypass the city, you should be able to navigate around it. Like all models, a road map is a simplified version of reality. If all the information about the city were on the map, you would be hopelessly confused by the excessive detail.

PROBLEM
Think about three different types of maps of your own choosing. How do these maps differ regarding (a) the details they include and exclude, (b) the regions they model, (c) their ease of portability, and (d) their ease of interpretation? Maps are designed for different purposes. How do your answers reflect the different purposes for which each type of map was designed?

All models change. In some cases, as knowledge about a subject increases, the models are made more complex in order to give a more accurate description of reality. In some cases, the process or structure that is being modeled changes, and the model must be modified to reflect the new reality. For example, when a trout stream that ran through New York City was diverted to build Wall Street, the blue line on the city map was replaced with a red line. Finally, the original purpose of the model may become irrelevant, and the model may be either abandoned or changed to reflect some new purpose.

Models of evolution are like maps; they represent the world in a simplified form. Maps allow people to get from one geographical point to another and to predict the efficiency of different potential routes. Likewise, evolutionary models allow biologists to see how populations go from one point in evolutionary history to another and to predict the route that a population may follow as it changes.

The Hardy-Weinberg model was created to solve two specific problems in genetics. Later, J. B. S. Haldane and other researchers modified the model to answer a variety of questions about evolution. Today this model is being applied by forensic scientists to assess the uniqueness of an individual's DNA fingerprint.

A POPULATION GENETICS PUZZLE

R. C. Punnett, one of the founders of mammalian genetics and originator of the Punnett Square, prompted the development of the Hardy-Weinberg model. Following one of his lectures, Punnett was asked: "If brown eyes are dominant over blue eyes, why doesn't a population eventually become all brown-eyed?" Punnett was baffled by this question, as were most biologists during the early days of genetics. The term *dominant* brings to mind a powerful trait that should overwhelm a weaker, "recessive" trait. Shouldn't the dominant trait always increase within the population?

Punnett mentioned this problem to his good friend Godfrey H. Hardy, one of the world's greatest mathematicians. Hardy and Punnett both worked at Cambridge University and had become acquainted by playing and watching cricket matches together. The next day Hardy rewarded Punnett with an answer that was published in the widely read journal *Science* in 1908.

Unknown to both Punnett and Hardy, the German physician Wilhelm Weinberg had developed a similar model six months earlier to help him investigate the question of whether there is a genetic basis to giving birth to twins. Weinberg published his findings in an obscure journal, and therefore scientists ignored his version of the model for another 35 years.

HARDY'S SOLUTION TO PUNNETT'S PUZZLE

Let's return to the map and use it as a model to guide you through a large city that is on the way to a summer vacation paradise (Figure 15.1). You pull out your map and have the choice of taking either I-5 or Route 99 through the city. Which route do you take? What factors do you use to make this decision? What information is on the map that helps you make this choice? You might notice that some features of interest for making this decision are missing from this model. For example, there is no information about traffic lights, mountains, and good places to eat. This particular model makes an assumption that these missing features are not important to the map reader.

Like the city map example, all models contain **assumptions**. To understand the Hardy-Weinberg model, we need to first understand its assumptions. For sexually reproducing organisms, Hardy assumed that all individuals of a population have an equal probability of combining their gametes (random mating). Though gametes have many thousands of genes, Hardy's model looked at only one gene, which was present in the population in two alternate forms, or **alleles**.

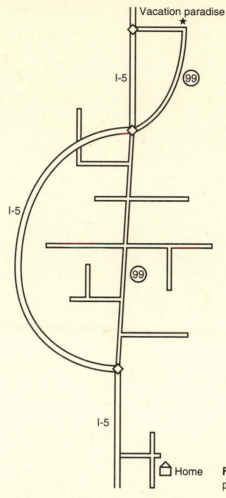

Vacation paradise

I-5

99

I-5

I-5

99

I-5

⌂ Home

FIGURE 15.1 Road map from home to vacation paradise.

If the brown-eye trait is dominant over the blue-eye trait, *EE* and *Ee* people will have brown eyes, while *ee* people will have blue eyes. Let's begin with a population made up of 9 homozygous dominant individuals (genotype *EE*), 42 heterozygous individuals (genotype *Ee*), and 49 homozygous recessive individuals (genotype *ee*). The total number of *E* and *e* alleles in the population is represented in the table below.

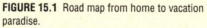

Genotype	Number of Individuals	Number of Copies of *E* Allele	Number of Copies of *e* Allele
EE	9	18	
Ee	42	42	42
ee	49		98
Total	100	60	140

The genotype frequency is the number of individuals of a particular genotype, divided by the total number of individuals in the population. Thus the initial genotype frequencies of this population are:

$$EE = 9/100 = 0.09, \; Ee = 42/100 = 0.42, \; ee = 49/100 = 0.49.$$

These frequencies can also be thought of as percentages, with 9 percent *EE*, 42 percent *Ee*, and 49 percent *ee*. Note that they then add up to 100 percent.

The gene pool is made up of 200 alleles because each individual has two alleles. Therefore the frequency of the *E* allele = 60/200 = 0.30, and the frequency of the *e* allele = 140/200 = 0.70. Just as mapmakers use dots and stars as symbols to represent cities, geneticists use the symbols *p* and *q* to represent allele frequencies, with *p* = the frequency of the dominant allele and *q* = the frequency of the recessive allele. In this case, *p* = 0.30 (or 30 percent) and *q* = 0.70 (or 70 percent). Because this population only has two alleles, *p* + *q* = 1 (or 100 percent).

Using these allele frequencies, Hardy asked: What will the genotype frequencies be in the next generation, assuming random mating? Will they remain the same, or will the brown-eyed allele increase in frequency? You can calculate the genotype frequencies for the next generation using the table below.

Sperm	Eggs	
	E = 0.30	*e* = 0.70
E = 0.30	*EE* = 0.09 (0.30 X 0.30)	*Ee* = 0.21 (0.30 X 0.70)
e = 0.70	*eE* = 0.21 (0.70 X 0.30)	*ee* = 0.49 (0.70 X 0.70)

According to Hardy's model, we can predict that the genotype frequencies in the next generation will be:

$$EE = 0.09$$
$$Ee = 0.21 + 0.21 = 0.42$$
$$ee = 0.49$$

These predicted frequencies are identical to the starting frequencies, and, as Hardy demonstrated, will always remain the same so long as the model's assumptions are not violated.

Let's see if the allele frequency has changed.

p = frequency of *EE* + 1/2 (frequency of *Ee*) = 0.09 + 1/2 (0.42) = 0.30
q = frequency of *ee* + 1/2 (frequency of *Ee*) = 0.49 + 1/2 (0.42) = 0.70

In Hardy's population, if mating is random, the allele frequency also remains constant from generation to generation. Note that there is no tendency for a dominant trait to increase in frequency from one generation to the next. Thus Hardy provided his friend Punnett with an answer to his question of why the dominant allele doesn't increase from generation to generation.

The Role of Assumptions in Modeling

In order to be useful, a model must make simplifying assumptions. For example, mapmakers assume that users are not interested in a three-dimensional scale model of all the buildings in a town. Such a model, though wonderfully accurate, would have difficulty fitting into the typical glove compartment. The Hardy-Weinberg model simplifies reality by assuming random mating. The other major assumption is that the four evolutionary factors of natural selection, gene flow, genetic drift, and mutation are not operating. This is an imposing list of assumptions, and on the surface you might wonder why we even bother with this model, given that it applies to no real population. There are three important reasons why this particular model is important, despite its unrealistic assumptions.

First, the general prediction of the Hardy-Weinberg model is true, even if we violate its assumptions to some degree. This is an important characteristic, because mating is never completely random, and evolutionary factors will be affecting allele frequencies to some degree in most populations. In the case of the Hardy-Weinberg model, even if the assumptions are not met, the genotype and allele frequencies will be similar (though not exactly identical) from one generation to the next.

Second, this model can be used as a diagnostic tool. Just as a physician bangs your knee with a hammer to test your reflexes, population geneticists use the Hardy-Weinberg model to determine if its assumptions hold for a particular population. If the genotype frequencies are changing, they know that the population has violated at least one assumption of the model in a major way. Then they can look for evidence of nonrandom mating or for evolutionary factors such as natural selection or genetic drift that may be operating in the population.

Finally, the history of this model shows that ways of thinking that develop for one purpose may often be applied for many other purposes. After the Hardy-Weinberg model was developed, it sat idle for over a decade, having successfully resolved the confusion over why dominant traits don't increase in frequency within a population. After World War I, however, the theoretical biologists R. A. Fisher, J. B. S. Haldane, and Sewell Wright built upon the model, ultimately providing the foundation for the new field of population genetics.

JOHN BURDON SANDERSON HALDANE

J. B. S. Haldane (Figure 15.2) had a unique scientific background. His father, a world authority on respiratory physiology, devoted a significant portion of his professional activities to investigating atmospheric conditions in mines. Haldane routinely accompanied his father on these investigations, and was once required by his father to stand up in an alcove of a mine shaft and recite Mark Antony's famous "Friends, Romans, countrymen" speech from *Julius Caesar*. After a few sentences, Haldane passed out

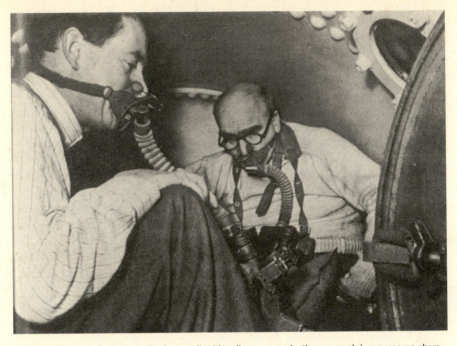

FIGURE 15.2 J. B. S. Haldane (background) with colleague, conducting research in a pressure chamber. *Source:* Ronald W. Clark, *JBS: The Life and Work of J.B.S. Haldane,* New York: Coward-McCann, Inc., 1969.

from inhaling methane that had collected at the ceiling of the alcove, effectively learning that methane is lighter than air and generally does not kill when inhaled. Such experiential learning formed a significant component of his informal education.

A second important episode in Haldane's informal education was World War I, during which he designed, tested, and lobbed experimental bombs at enemy artillery. Haldane relished these activities, earning for himself the nickname "Bombo" and a reputation as the bravest and dirtiest officer in the army. He used his respiratory physiology background to design a gas mask (using himself as the "guinea pig") to provide protection against poisonous gas attacks launched by the Germans.

This man of action who designed sensitive mechanisms for grenades and bombs and personally hurled his creations at the enemy contrasts with Haldane the laboratory scientist. With the exception of his specialized respiratory equipment, which he could use quite effectively, Haldane was very clumsy when it came to operating delicate equipment or handling small specimens for his genetics research. We don't know whether this ineptitude guided his research agenda at all, but most of Haldane's work was done with his brain, a pencil and paper, and the language of mathematics.

Haldane was fascinated by the fruit fly research of T. H. Morgan and H. J. Muller, experiencing "all the satisfaction of reading a first-rate detective story, much enhanced by the fact that the story was true" (see Chapter 5). He realized that he could help tie together the fields of genetics and evolution if he could derive a quantitative theory to show how natural selection changes allele frequencies. Haldane had two problems. First, he needed to develop a model that related the intensity of natural selection to the rate of evolutionary change. Second, he needed to apply his model to a species that had changed genetically in some clear and significant manner in the recent past.

Haldane knew that Hardy's model predicted no change in genotype frequency. He also knew that Hardy's model assumed that natural selection was not operating in the population. Haldane reasoned that one approach to developing his model would be to modify Hardy's model by including the effects of natural selection. But he needed a measure of natural selection's intensity.

Haldane defined fitness (w) as the relative ability of an organism to pass on its genes to the next generation. Fitness is affected by two factors: an individual's ability to survive to sexual maturity and an individual's reproductive success. Both of these depend on how well adapted the individual is to a particular environment. According to Haldane's definition, the most successful genotype has a fitness (w) = 1.0. A genotype that produces, on average, only 20 percent as many offspring as the most successful genotype has $w = 0.20$.

Haldane defined the selection coefficient (s) as a reduction of fitness suffered by the less successful genotype; it is equal to $1 - w$. In this example, the more successful genotype has no reduction in fitness, so $s = 0$, and the less successful genotype has an 80 percent reduction in fitness, so $s = 1.0 - 0.20 = 0.80$.

One of the best-known populations that had changed genetically in Haldane's recent past was the peppered moth (see Chapter 1). Between 1848 and 1901, the population near Manchester, England, went from almost 100 percent light-winged moths to almost 100 percent dark-winged moths. Haldane suspected that natural selection was responsible for this change, and he used this example to calculate how strong natural selection must be to bring about such a dramatic change in genotype frequency in only 53 generations.

PROBLEM
Propose at least two different factors that could give a selective advantage to the dark-winged moths in Manchester.

By incorporating the selection coefficient into the basic Hardy-Weinberg equation, Haldane was able to show how a population would change in cases where selection was operating against either the recessive allele or the dominant allele. Haldane knew that light wing color was due to a recessive allele in the peppered moths. According to his new formula, the change in the frequency of a recessive allele (Δq) in one generation is $\Delta q = -spq^2/(1 - sq^2)$.

Haldane assumed there were a few dark-colored moths in Manchester in 1848, and he began with the frequency of the dark-winged moths = 0.01 and the frequency of the light-winged moths = 0.99. He then plugged in different possible values for the selection coefficient, repeating this process 53 times (without a calculator) to represent the 53 generations between 1848 and 1901. For each generation, the allele frequencies changed in response to the intensity of natural selection. Haldane estimated that an average dark-winged moth must produce three surviving offspring for each two offspring produced by the average light-winged moth in the Manchester population.

PROBLEM
Based on Haldane's estimates of rates for the two forms of the peppered moth, what are the values of *w* and *s* for the dark- and light-winged forms of the peppered moth?

Approximately 30 years later, H. B. D. Kettlewell conducted field experiments that measured the selective advantage to the dark-winged moths in Manchester. While there were some differences between Haldane's estimation and Kettlewell's field measurement, the two values were reasonably close. This demonstrated that Haldane's model of evolution could make testable predictions about evolution. During the 1920s and 1930s, Haldane, Fisher, and Wright also modified the basic model to show how genetic drift, gene flow, and mutation could bring about evolutionary change. Ultimately, these modeling efforts tied together the fields of genetics and evolutionary biology.

☐ EPILOGUE

One of the reasons for focusing discussion on the Hardy-Weinberg model is that it provides a good example of how a model may be modified and applied to solve different problems. The Hardy-Weinberg model is now being applied by forensic scientists to estimate the uniqueness of DNA fingerprints.

In one notorious example, an assailant broke into the house of a computer operator in Orlando, Florida. He covered the victim's face with a sheet and raped her. This same pattern was repeated over the next year, with police suspecting the same man to be responsible for 23 incidents of breaking and entering and attempted assault or rape. On March 1, 1987, police apprehended and arrested a suspected prowler.

The original victim had seen the suspect before her face was covered, so she was called to examine a photo lineup. She immediately identified the suspect from the group of photos presented to her. A problem arose because none of the other victims was able to make a visual identification, and the suspect had two witnesses who swore he never left the house at the time of the crimes.

Fortunately, police had collected sperm samples from the original victim immediately after she reported the crime. The DNA pattern of the rapist's sperm was identical to the DNA from the suspect's white blood cells. The testing laboratory estimated the odds of such a match as approximately 1 in 10 billion. How did the lab make this numerical claim?

Forensic scientists can make such estimates because there are regions of DNA that don't appear to code for any types of protein. Some of these regions are made up of repetitive DNA—regions of DNA that repeat the same base-pair sequences over and over again. Molecular biologists have identified regions that are highly variable in the number of repeating sequences. This type of gene is called a VNTR, for "variable number of tandem repeats." Within a population, there may be hundreds of alleles of a particular VNTR gene, with each allele representing a different number of tandem repeats (Figure 15.3).

(A) TGTTTATGTTTATGTTTATGTTTA

(B) ==⇒==
    ==⇒⇒==
    ==⇒⇒⇒==
    ==⇒⇒⇒⇒==
    ==⇒⇒⇒⇒⇒==

FIGURE 15.3 (A) Repetitive sequence of one strand of DNA, made up of four tandem repeats of the six-base sequence TGTTTA. (B) Five alleles of VNTR genes that differ in number of tandem repeats (= represents nonrepetitive part of the DNA; ⇒ represents tandem repeats). Notice that there are other parts of the genes besides the tandem repeats. Also notice that the genes with more tandem repeats are larger. This size difference allows each allele to be distinguished by forensic scientists.

Because there are so many different VNTR alleles, the frequency of each individual allele is very low. If the frequencies of the two alleles are 0.01 and 0.02, then according to the Hardy-Weinberg model, the genotype frequency of a heterozygote individual will be $2pq$ or 2(0.01)(0.02) or 0.0004. On average, only 4 out of 10,000 individuals in the human population would have that genotype. In this way, the allele frequencies from a large random sample of the population form a baseline from which the expected genotype frequencies are calculated.

Furthermore, forensic scientists can simultaneously look at more than one VNTR gene for a given individual. By assuming that a VNTR genotype at one gene does not affect the probability of having a particular VNTR genotype at a second gene, scientists are able to more effectively discriminate between DNA samples. The assumption that genotype probabilities at two (or more) different genes are unrelated is an example of the assumption of statistical independence and only holds in a population that is in Hardy-Weinberg equilibrium. Under these conditions, the probability of having two matching genotypes is the product of the probability of having each of those genotypes independently (this is called the product rule).

Returning to the previous example, the genotype frequency at the first VNTR gene was 0.0004. Let's consider a second VNTR gene for the same individual, who is heterozygous for two alleles that have frequencies of 0.03 and 0.04. The genotype frequency is $2pq$ = 2(0.03)(0.04) = 0.0024. According to the product rule, the probability of having both of those genotypes is the product of the genotype frequencies within the population = 0.0004 x 0.0024 = 0.00000096. Only 96 individuals would be expected to have that genotype in a population of 100 million.

Most forensic scientists look at four different independent VNTR genes for criminal cases. They believe that this number of genes is sufficient to rule out the possibility of a false conviction. Other geneticists disagree, pointing out that the product rule may not be valid for human populations. If random mating does not occur, then there may be a higher probability that two individuals will, by chance, have the same VNTR genes. Recently, a panel of scientists from the National Academy of Sciences endorsed a more conservative type of analysis that accounts for possible nonrandom mating and gives somewhat lower probabilities of uniqueness of DNA fingerprints. Though this debate continues, the Hardy-Weinberg model lies at the heart of this new technology, bringing criminals to justice and clearing innocent suspects of wrongdoing.

QUESTIONS AND ACTIVITIES

1. What does this case show about the following aspects of doing biology?
   — different uses of models
   — the assumptions made by models
   — how models change over time

2. Why is high fitness not the same as dominance? How does understanding this distinction allow you to answer the question posed to Punnett after his lecture?

3. In this chapter, Haldane predicted the selective advantage of the dark-winged form of the peppered moth in industrialized areas, based on how long it took the dark-winged form to spread within a population. As described in Chapter 1, Kettlewell measured the selective advantage in the field. Are the numbers in close agreement? What might account for any differences between Haldane's predicted values and Kettlewell's findings?

4. One of the characteristics of the Hardy-Weinberg model is that its predictions hold even when its assumptions are not completely met. As a prosecuting attorney, tell the jury why this characteristic is important for them to understand when considering DNA fingerprinting evidence.

5. Haldane made a model of evolution in the case where selection operated against a dominant allele as well as in the case of selection against the homozygous recessive allele described above. In which case would you expect evolution to proceed more rapidly? Why?

6. The initial population in the eye-color example in the text was already in Hardy-Weinberg equilibrium. Hardy's original *Science* paper began with a population made up exclusively of homozygous individuals; thus, his population was not in equilibrium. Calculate the allele frequency and the genotype frequencies of the next generation for a population consisting of 30 homozygous brown-eyed individuals and 70 homozygous blue-eyed individuals. How do these frequencies compare with the eye-color example in the text? If a population is disturbed from Hardy-Weinberg equilibrium, how long does it take to return to equilibrium genotype frequencies?

7. Revisit the example from the text where $w = 0.20$ for the less successful form (remember w always equals 1.0 for the most successful form). Realize that this is an example of very intense natural selection, as the most fit genotype is producing five times as many offspring as the less fit genotype. If $w = 0.20$, then $s = 1 - 0.20 = 0.80$. Let's begin with an intermediate frequency of the recessive allele of $q = 0.60$ and see how the allele frequency changes from one generation to the next. The change in the frequency of the recessive allele, $\Delta q = -spq^2/(1 - sq^2)$, $= -0.80(0.40)(0.60)^2/(1 - 0.80(0.60))^2 = -0.1152/(1 - 0.288) = -0.1618$. Thus q for the next generation is $0.60 - 0.1618 = 0.4382$. This dramatic decrease in the frequency of the recessive allele occurs because the fitness of the homozygous recessive individual is so low.

Intuitively, do you think evolutionary change will be most rapid at high, intermediate (as in our example), or low frequencies of the recessive allele? Haldane was particularly interested in how fast evolutionary change should occur under different conditions. Help him solve this problem, and test your intuition by using his formula to determine if Δq is greatest at high, intermediate, or low frequencies of the recessive allele.

SUGGESTED READING

Clark, R. W. 1969. *JBS: The Life and Work of J. B. S. Haldane*. New York: Coward-McCann.

Hardy, G. H. 1908. "Mendelian Proportions in a Mixed Population." *Science* 28: 49–50.

Judson, H. F. 1987. "Modeling." In *The Search for Solutions*. Baltimore: Johns Hopkins University Press, pp. 111–129.

Levins, R. 1966. "The Strategy of Model Building in Population Biology." *American Scientist* 54: 421–431.

Lewis, R. 1988. "DNA Fingerprints: Witness for the Prosecution." *Discover* (June): 45–52.

Punnett, R. C. 1950. "Early Days of Genetics." *Heredity* 4: 1–10.

Weinberg, W. 1908. "On the Demonstration of Heredity in Man." In S. H. Boyer, IV, ed., *Papers on Human Genetics*. Englewood Cliffs, NJ: Prentice-Hall, pp. 4–15. (Reprinted from *Jahreshefte des Vereins für Vaterländische Naturkunde in Württemberg, Stüttgart*. 1908. 64: 368–382. Lecture at the scientific evening at Stüttgart, January 13, 1908.)

CHAPTER 16

George Gaylord Simpson *&* the Question of Continental Drift

JOEL B. HAGEN

☐ *INTRODUCTION*

When Europeans began worldwide exploration in the 1500s, they were confronted with a wealth of biological information. The discovery of exotic species from new lands raised troubling questions. Why are there so many species? Why do countries with similar climates often have different species of plants and animals? Did all species originate in one spot? If so, how could they migrate to distant parts of the world?

It is easy to chuckle at early attempts to answer these questions. Faced with the problem of accommodating hundreds of new species, some scientists painstakingly recreated the floor plan of Noah's ark. Others debated the problems faced by penguins and reindeer migrating back to their polar homes after the Flood. However unscientific these efforts may strike us, they were the first attempts to explain the geography of plants and animals. This early biogeographical tradition provided a background for the evolutionary theories proposed by later naturalists such as Charles Darwin and Alfred Russel Wallace.

Biogeography provided such crucial evidence for evolution that Darwin devoted two chapters to the topic in his book *On the Origin of Species*. When he had visited the Galapagos Islands as a young man, Darwin was amazed to find that although the birds, tortoises, and lizards were similar to those found in South America, the species were not the same. In fact, some islands had unique species found nowhere else in the world. Darwin reasoned that all of these species had evolved from a few South American ancestors accidently carried to the islands by storms or ocean currents. Because the islands are somewhat isolated from one another and their environmental conditions are not the same, a new species might evolve on one island, but not on others.

This type of evolutionary explanation revolutionized biogeography, but it did not provide ready-made answers to such questions as why kangaroos and their diverse relatives (marsupials) are common in Australia and the New World continents but absent from Europe, Africa, and Asia (see Figure 16.1(A)). One of the twentieth-

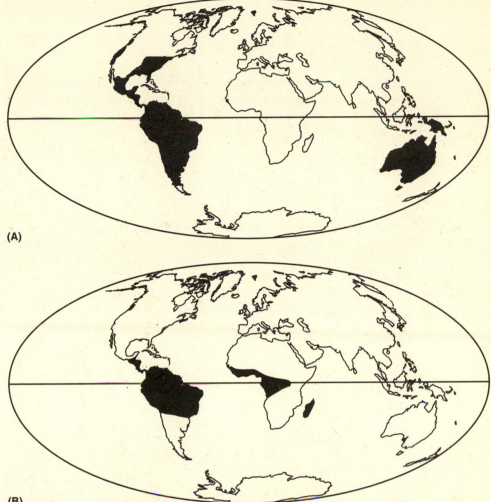

FIGURE 16.1 (A) The discontinuous distribution of the order Marsupialia, which includes kangaroos, koalas, opossums, wallabies, and many other pouched mammals. (B) The discontinuous distribution of species in the genus *Symphonia*. This genus of small shrubs includes 2 species found only in South America, 1 species found both in South America and Africa, and 18 species restricted to the island of Madagascar.

century biogeographers who tried to answer this question was George Gaylord Simpson, a paleontologist and leading authority on the evolution of mammals.

To explain such unusual patterns of distribution, Simpson and other biogeographers had to consider a wide variety of factors including climate, habitat, competition, dispersal, speciation, extinction, and geographic barriers. Although these factors are partly biological, they are also heavily influenced by the physical environment.

Biogeographers, therefore, had to borrow theories and data from other sciences, particularly geology. As a result, they were drawn into one of the great interdisciplinary controversies in twentieth-century science: the question of continental drift.

PROBLEM

Describe several plausible explanations for the two biogeographical patterns illustrated in Figure 16.1(A) and (B). In formulating each of your hypotheses, consider the following factors: climate, habitat, competition, dispersal or migration, speciation, extinction, and geographic barriers.

A GEOLOGICAL CONTROVERSY

Spanning the half century from the 1920s through the 1970s, Simpson's scientific career coincided with a major controversy about the nature of the earth. The controversy was sparked by the publication of a short book written during World War I by the German scientist Alfred Wegener. In his book, Wegener claimed that all continents had once been joined in a single landmass that he called Gondwanaland. Over the course of millions of years, this supercontinent disintegrated into several pieces that gradually drifted apart, forming the continents of today.

Although similar claims had been made before, Wegener was the first scientist to gather a large body of evidence to support **continental drift**. Some of his most controversial evidence came from biogeography. For example, Wegener claimed that closely related species of earthworms are found on either side of the Atlantic Ocean. Obviously, their ancestors could not swim across the ocean or burrow through the polar ice, so how did earthworms become so widely distributed? According to Wegener, continental drift provided the most plausible explanation.

Most biologists rejected Wegener's theory. He was not a biologist, and many biogeographers dismissed him as an amateur dabbling in a field that he did not understand. His critics also pointed out that Wegener could not adequately explain how continental drift worked. He claimed that the gravitational attraction of the sun and moon pulled the continents apart, but this mechanism was almost universally dismissed by physicists and geologists. How could continents push through the solid, rocky crust of the earth? What force was sufficiently powerful to drive this movement? It seemed about as likely as pushing this book through a concrete wall. Without a convincing mechanism for continental drift, most biogeographers rejected the theory as a wild flight of fancy.

PROBLEM

Reconsider your explanations for the patterns of distribution illustrated in Figure 16.1. Would Wegener's theory help you to explain these biogeographical patterns? If so, how would you use it in each example? Before you could explain these biogeographical patterns, why would you also need to know *when* Gondwanaland broke up and *how long* it took for the new continents to drift apart?

If continents did not move, perhaps they had once been connected by **land bridges**. For example, if Brazil and Africa had been connected by such a passageway, plants and animals could have freely migrated between the continents (Figure

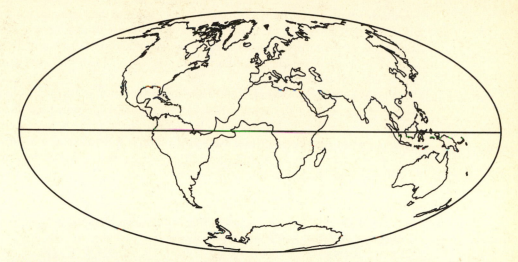

FIGURE 16.2 Hypothetical land bridge connecting South America and Africa. As late as World War II, some biogeographers used land bridges to explain the global distributions of plants and animals.

16.2). Later, if the bridge had collapsed, isolated populations on either side of the Atlantic Ocean could have evolved independently. This was a convenient theory for biogeographers, and during the late nineteenth century they postulated numerous land bridges joining the continents.

What might have caused these massive bridges to disappear? Most geologists in the 1800s believed that the earth originated in a molten state. As it cooled and shrank, tremendous stresses built up in the thin, rocky crust surrounding the still-molten core. Periodically, these stresses were released by the catastrophic collapse of parts of the crust. During these cataclysms, some geologists believed, old land bridges sunk into the ocean floors and new ones were sometimes formed.

Land bridges had been popular among biogeographers, but by the time Wegener wrote his book, the theory was losing support. The discovery of radioactivity around 1900 challenged the assumption of a cooling earth. Perhaps radioactivity deep in the earth's core generated a uniform output of heat. Rather than cooling, the earth's temperature might actually have remained fairly constant. If so, there would be no contraction to cause the earth's crust to buckle and collapse. And was there any evidence for land bridges? If they once existed, the remnants should still be found on ocean floors. Despite efforts to find them, however, the hypothetical land bridges seemed to have disappeared without a trace.

PROBLEM

Reconsider your explanations for the patterns of distribution illustrated in Figure 16.1. How would land bridges help you to explain these biogeographical patterns? Explain whether a biogeographer would be justified in accepting hypothetical land bridges without evidence for their prior existence.

Continental drift and land bridges attracted a few supporters after World War I, but most biogeographers, including Simpson, accepted a third geological theory:

permanentism. According to this theory, oceans and continents have always existed where they are today. Because there was no compelling evidence that the sizes, shapes, or relative positions of the continents had changed substantially, most geologists felt justified in believing that the surface of the earth had always been pretty much as it is today.

Permanentists acknowledged that many smaller geological changes have occurred throughout the earth's history. Continental glaciers form and retreat in response to climatic changes. Sea levels fluctuate up and down. Mountain ranges are thrust upward and then erode. Islands are formed by volcanic action or by the growth of coral reefs. Compared to drifting continents or collapsing land bridges, however, these changes are minor.

PROBLEM
Reconsider your explanations for the patterns of distribution illustrated in Figure 16.1. To what extent are they consistent with permanentism?

The situation faced by biogeographers from about 1920 to 1970 is not unusual in science. Scientists are usually trained as specialists—experts in rather narrow areas of research. Yet solving scientific problems often requires theories, data, or instruments borrowed from other fields. In this case, biogeographers needed a geological theory, but most of them were not experts in geology. The geologists who they turned to for guidance strongly disagreed over the three alternative theories. On what rational basis could a biogeographer, such as Simpson, choose among the competing geological theories?

To some extent, Simpson's choice might be influenced by teachers or other scientists. Some of Simpson's teachers were prominent opponents of continental drift. American scientists were almost unanimous in supporting permanentism, and very few American biogeographers argued for land bridges or continental drift. In Europe, the situation was slightly different. There, continental drift and land bridges were also controversial, but they did have some prominent supporters.

A biogeographer might also choose a theory based on the type of organism studied. For example, after 1920 a few European botanists cautiously accepted continental drift. They needed to explain how closely related species of plants were sometimes separated by vast expanses of ocean (see Figure 16.1(B)). In many cases it was hard to imagine how plants could disperse over such large geographic areas. Seeds are quickly killed by seawater, so it seemed unlikely that they floated from one continent to another. Migratory birds might carry a few seeds, but only along well-established flyways. On the other hand, if the continents had moved, then many troubling patterns of distribution could be explained. Despite its unpopularity, these botanists were willing to accept the intellectual risk of using continental drift as a tentative "working hypothesis." Although they admitted that it was speculative, they hoped that one day geologists would successfully explain how the process worked.

Like Simpson, most zoologists did not believe in drifting continents or massive land bridges. They knew that, over the course of millions of years, even slow-moving animals can disperse very widely. Tiny insects and other small animals can sometimes be carried for hundreds of miles by wind. Darwin and other early biogeographers had also pointed out that animals floating on driftwood rafts occasionally are carried very long distances by ocean currents. Such movements could explain many unusual patterns of animal distribution. Not surprisingly, almost all zoologists were strong supporters of permanentism. It required fewer assumptions than the competing theories, and it appealed to common sense. The alternative theories required a belief in unproved—perhaps unprovable—geological forces of enormous magnitudes. If geologists were skeptical of these speculative theories, why should zoologists seriously consider land bridges or continental drift?

A PERMANENTIST GEOGRAPHY OF ANIMALS

Like many paleontologists, Simpson (Figure 16.3) was trained as a geologist. As a student at Yale University, however, he also took many biology courses. He developed a keen interest in heredity and was strongly influenced by theories of Darwinian natural selection and Mendelian genetics (Chapter 15). As a scientist at the American Museum of Natural History in New York, Simpson became an authority on the evolution of mammals. He wrote over 700 books and articles, including several on evolutionary theory aimed at general audiences. His views on the geography of animals, including his criticisms of continental drift and land bridges, were widely read and highly influential.

FIGURE 16.3 George Gaylord Simpson, paleontologist, biogeographer, and leading critic of continental drift. *Source:* Simpson, G. G. 1965. *Concession to the Improbable: An Unconventional Autobiography.* New Haven, CT: Yale University Press.

Although he usually discussed mammals, Simpson believed that the same basic principles could explain the geographical distribution of all animals and even plants. When climatic conditions are favorable, species migrate quite freely across large expanses of land called **corridors**. The broad connection between Europe and Asia is a good example of a corridor uninterrupted by an ocean or other major geological barriers.

Filter bridges are temporary connections between continents. Such bridges include the Isthmus of Panama between North and South America and the chain of Aleutian Islands in the Bering Sea between Asia and North America. At some times during the earth's history, these bridges have been completely above water, and at other times they have been submerged. As the term implies, filter bridges allow some species to migrate but restrict the movements of others. For example, many mammals from temperate regions crossed the filter bridge between Asia and North America, but the Bering region was too cold for tropical mammals to successfully make the journey. By restricting dispersal, filter bridges periodically allow populations to evolve in isolation from one another.

Sweepstakes routes are ocean currents that carry animals on logs or other floating debris from island to island ("island hopping") or from one continent to another. As the name implies, sweepstakes routes involve a large element of chance. Although rarely traveled, these routes can have major evolutionary consequences when rafters "hit the jackpot." A single pregnant female landing on an isolated island might serve as the starting point for the evolution of a unique and diverse fauna.

Perhaps the best example of this type of random, sweepstakes dispersal is found on the Hawaiian Islands. Because the islands are only a few million years old and are isolated in all directions by hundreds of miles of ocean, it is unlikely that they were ever connected to any of the continents. Therefore, the ancestors of all island species must have floated across the ocean. Hawaii has no native mammals which apparently never crossed the sweepstakes route. But over 3,700 unique species of insects live on the islands. Claiming that these insects evolved from about 250 ancestral species, Simpson estimated that, on average, an insect successfully crossed a sweepstakes route to Hawaii about once every 20,000 years.

PROBLEM
Why might insects be much more successful than mammals at crossing sweepstakes routes like the ones leading to the Hawaiian islands?

To understand how Simpson used these ideas to explain evolution and geographical distribution, consider the marsupials (see Figure 16.1(A)), which diverged from placental mammals perhaps 100 million years ago. For about 30 million years, marsupials competed on even terms with primitive placentals.

Modern placentals evolved about 65 million years ago and drove most of the earlier groups of mammals to extinction. Marsupials flourished in only two remote geographical refuges: Australia and South America.

Simpson combined evidence, logic, and speculation to recreate this evolutionary history of the marsupials. Comparing the characteristics of living species suggested that the ancestral marsupial was similar to an opossum. For example, opossums have relatively unspecialized, general-purpose limbs, compared to the highly specialized, jumping hind legs of kangaroos. It seemed logical that a specialized leg evolved from a more generalized leg, rather than vice versa. Other anatomical comparisons among living and fossil marsupials also supported this hypothesis. Because the oldest opossum fossils had been found in North America, Simpson concluded that this was probably the area where marsupials originated.

Abundant fossils provided strong evidence that marsupials migrated to South America soon after they originated. When the filter bridge at the Isthmus of Panama submerged, the marsupials—and also some primitive placentals—were protected from competition with the more modern placentals which began to evolve about 65 million years ago. Marsupials flourished and diversified in this isolated habitat until the filter bridge was reestablished about 2 million years ago. Although most of the pouched mammals did not survive the onslaught of better-adapted placentals which migrated from the north, several groups of well-adapted marsupials persisted in South America.

Simpson who was an expert on South American mammals, provided a convincing explanation for how marsupials became widespread in South America. But how did the early marsupials disperse to the remote island continent of Australia?

Fossil evidence could not help Simpson answer this question because very few marsupial fossils had been found in Europe, and none had ever been found in Asia. Simpson reasoned, however, that because so many other ancient mammals had successfully crossed the Bering Strait, there was no reason to think that marsupials had failed to use this filter bridge connecting North America and Asia. If this were true, then it was only a matter of time before marsupial fossils would be discovered in Asia.

Once established in Asia, Simpson believed that some ancient marsupials had island hopped across a sweepstakes route from southeast Asia to Australia. Was Simpson justified in proposing this speculative hypothesis without direct fossil evidence to support it? In response, he argued that compared to the alternative hypotheses, his was the most reasonable explanation.

Simpson denied that land bridges were involved in the diversification and spread of marsupials. Some earlier zoologists had claimed that marsupials migrated from southeast Asia to Australia across a land bridge. Yet this hypothesis could not easily explain why placental mammals failed to invade Australia. Furthermore, Simpson pointed out, there was no convincing evidence that such a land bridge had ever existed. It seemed more logical to hypothesize a sweepstakes route that just by chance allowed marsupials, but not placentals, to reach Australia.

Simpson also dismissed Wegener's claim that Australia, Antarctica, and South America had once formed a single continent. Although this might explain why marsupials are so common in both Australia and South America, continental drift was too speculative to take seriously. Echoing Simpson's view, a popular textbook author in the 1950s bluntly concluded that the present distribution of mammals was a product of moving animals, not moving continents.

PERMANENTISM ADRIFT

During the 1950s, when Simpson's biogeographical writings had their greatest influence, the theory of permanentism seemed as rock solid as the apparently unmovable continents. Twenty years later, both the theory and the continents were adrift.

Geologists and oceanographers were gathering convincing evidence that the continents were indeed moving, although not in the way that Alfred Wegener had claimed. According to the new **theory of plate tectonics**, the crust of the earth is composed of several huge plates that float on the molten mantle below. Circulating currents of hot magma in the mantle push the plates in various directions. In certain parts of the world (subduction zones), the leading edge of one plate dips beneath its neighbor. At the trailing edge of the plate, new crust is formed by upwelling magma from the earth's mantle. The continents sit atop the huge crustal plates and move along with them. Two continents on adjacent plates will drift together or apart depending on the movements of the underlying plates.

Plate tectonics revolutionized the way geologists thought about the earth, and it also had important implications for biogeography. It now seemed likely that the shapes and relative positions of continents had been in a constant state of flux. Like huge arks, the drifting continents carried populations of plants and animals—sometimes joining, sometimes isolating the floras and faunas. This could profoundly influence both evolution and dispersal.

The revolution in geology placed Simpson and most other biogeographers in a difficult position. For over 20 years he had been a leading critic of continental drift. His views, published in many books, had been widely accepted by other biologists. Although he was in his late sixties, Simpson continued to write about evolution and biogeography. He admitted that the evidence for continental drift was overwhelming and that it was an important factor in animal distribution. For example, it now appeared that Alfred Wegener had been correct in claiming that Australia, Antarctica, and South America once formed a single continent, or at least had been neighboring islands. Simpson's earlier claim that marsupials had migrated from Asia to Australia now seemed unlikely. Rather than assuming that marsupial fossils would one day be found in Asia, he now thought it more likely that marsupials had never been on the continent.

In retrospect, it would be easy to conclude that Simpson and most other biogeographers had chosen the wrong geological theory. But was he "wrong" during the 1940s and 1950s when he rejected continental drift? Simpson later defended his choice by pointing out important differences between Wegener's early theory of continental drift and the newer theory of plate tectonics. Prior to the rise of plate tectonics, there was no convincing mechanism for continental movement. Geologists and physicists could not agree whether such movement was even theoretically possible. The geographical distribution of animals, and particularly mammals, could be adequately explained without continental drift. Permanentism seemed to be the simplest, least speculative, and most logical of the three alternative geological theories. In the end, however, it also turned out to be untenable.

Continental drift, once so unpopular, came to play a critical role in explaining biogeography. As Simpson later pointed out, however, it didn't automatically solve biogeographical problems. For example, knowing that South America, Antarctica, and Australia had once been neighboring islands did not fully explain marsupial migration patterns. Where had the marsupials originated? In what direction or directions had they moved? When did the movements occur? These questions could not be adequately answered without continental drift, but answering them also required some of Simpson's data and ideas. Corridors, filter bridges, and sweepstakes routes remained important parts of any biogeographical theory. The rise of plate tectonics was a great scientific revolution. Like all revolutions, however, it did not completely overthrow the past.

☐ *EPILOGUE*

Almost all living species of marsupials are found in Australia (120 species) and South America (80 species). Several species occur in Central America, but only 2 species (both opossums) are found in the United States. No marsupial species live in Europe, Asia, or Africa. This discontinuity continues to intrigue biologists.

In contrast to earlier theories, most zoologists now believe that marsupials were never distributed worldwide. According to the modern theory, marsupials arose in North America. Fossils suggest that a few early marsupials migrated to Europe, which was still connected to North America. But all marsupials in the Northern Hemisphere became extinct between 15 and 20 million years ago.

Long before the northern extinction occurred, a few marsupials rafted to South America, which was an island widely separated from North America and Africa. The tip of South America was close to another large island continent made up of what is now Antarctica and Australia. Although this continent was situated near the South Pole, it had a temperate climate. Recently discovered fossils prove that at different times during the Cretaceous Period (from about 145 to 65 million years ago), dinosaurs and mammals thrived on this ancient continent.

At the end of the Cretaceous Period, Australia separated from Antarctica and began drifting to its present location. Without competition from placental mammals, marsupials continued to diversify on this isolated island. The American continents drifted together about 2 million years ago, allowing modern placental mammals to migrate south across the Isthmus of Panama. This led to the extinction of many marsupial species that had previously flourished on the island of South America. Others, like the modern opossums, moved northward.

Although marsupials are less diverse today than in the past, it would be a mistake to dismiss them as poorly adapted, "second-class mammals." Consider the opossum, which is probably similar to some of the earliest marsupials. Since the colonial period, opossums have expanded their range about 500 miles northward through the eastern United States and into Canada. After being artificially introduced into California by settlers, these familiar marsupials rapidly spread throughout all of the Pacific states. The success of these highly adaptable marsupials should remind us that evolution is an ongoing process, and that many factors continually shape the geographical distribution of animals.

QUESTIONS AND ACTIVITIES

1. What does this case show about the following aspects of doing biology?
   — relationship between data and theories
   — expertise and interdisciplinary problems
   — revision of scientific theories
   — resolving scientific controversies

2. Simpson drew a sharp distinction between his filter bridges and the land bridges that earlier biogeographers had hypothesized (see Figure 16.2). Discuss the similarities and differences between the two ideas.

3. Many scientists believe that choices among competing theories should be made on the basis of *parsimony*. According to this principle, a scientist should always choose the simplest explanation or the one involving the fewest assumptions. Did Simpson follow this principle before 1970? Did he follow it after 1970? What about the botanists who used continental drift in the 1920s to explain the geographical distribution of plants?

4. Although Simpson was an expert on mammalian evolution, he believed that his biogeographical ideas (corridors, filter bridges, and sweepstakes routes) could be used to explain the distribution of all organisms (plants and animals). Are mammals typical of other organisms? What mammalian characteristics might make their migration patterns unusual?

SUGGESTED READING

Austad, S. N. 1988. "The Adaptable Opossum." *Scientific American* 258(2): 98-104.

Browne, J. 1983. *The Secular Ark: Studies in the History of Biogeography*. New Haven, CT: Yale University Press.

Colbert, E. H. 1985. *Wandering Lands and Animals: The Story of Continental Drift and Animal Populations*. New York: Dover.

Colbert, E. H., and M. Morales. 1991. *Evolution of the Vertebrates*. 4th ed. New York: John Wiley.

Frankel, H. 1981. "The Paleobiogeographical Debate over the Problem of Disjunctively Distributed Life Forms." *Studies in the History and Philosophy of Science* 12: 211-259.

Frankel, H. 1985. "The Biogeographical Aspect of the Debate over Continental Drift." *Earth Sciences History* 4(2): 160-181.

Le Grand, H. E. 1988. *Drifting Continents and Shifting Theories*. Cambridge, England: Cambridge University Press.

Simpson, G. G. 1965. *The Geography of Evolution: Collected Essays*. New York: Chilton.

Simpson, G. G. 1978. *Concession to the Improbable: An Unconventional Autobiography*. New Haven, CT: Yale University Press.

Simpson, G. G. 1980. *Splendid Isolation: The Curious History of South American Mammals*. New Haven, CT: Yale University Press.

CHAPTER 17

Rachel Carson *&* *Silent Spring*

DOUGLAS ALLCHIN

☐ **INTRODUCTION**

In the late 1950s, the United States enjoyed an economic prosperity that it had not experienced for three decades. Industry boomed after World War II. Comfortable middle-class neighborhoods—each house with its own lawn—began to expand into the suburbs of metropolitan areas. Scientific research, too, expanded. "Better living through chemistry" was the watchword of one company. Many chemical pesticides, such as DDT and aldrin, were used to control insect populations and thereby reduce disease. Farm production, too, boomed. New fertilizers and pesticides were widely available and allowed better crop yields. Farming was one of many activities indebted to the chemicals industry. The pesticides DDT and aldrin were emblems of the triumph of technology.

But amid the welcome prosperity, there were signs that all was not ideal. In residential communities, some people noticed that songbirds were declining. Birds seemed to be afflicted after heavy pesticide sprayings. One birdwatcher in Massachusetts, Olga Huckins, expressed her concerns to the editor of the Boston *Herald* in 1958:

> The mosquito control plane flew over our small town last summer. Since we live close to the marshes, we were treated to several lethal doses as the pilot criss-crossed our place…. The "harmless" shower bath killed seven of our lovely song-birds outright. We picked up three dead bodies the next morning right by our door…. All these birds died horribly, and in the same way. Their bills were gaping open, and their splayed claws were drawn up to their breasts in agony.

Her experience contradicted state officials, who had advised her that the spraying mixture—fuel oil with DDT—was "entirely harmless." She noted with irony that the grasshoppers and bees were gone, but not the mosquitoes themselves.

Mrs. Huckins was no idle birdwatcher, however. She asked an acquaintance, Rachel Carson (Figure 17.1), whom she might consult in the government. Carson had earned wide acclaim with two best-selling books popularizing marine biology. In these books, Carson evoked a fascination and respect for the ocean and its many crea-

FIGURE 17.1 Rachel Carson near her home in Maine while she was writing *Silent Spring. Source:* ©1962 Erich Hartmann/Magnum Photos Inc.

tures. She conveyed a sense of the delicate balance of nature. After a few inquiries, Carson realized that scientific experts knew much about the effects of pesticides, though the knowledge was not guiding policy decisions. A crisis seemed imminent. In part, Carson saw pesticides as a threat to the living things that she cherished and wrote about. She decided that she must write and inform the public about pesticides.

The challenge facing Carson was similar to that faced by many biologists: to marshall available evidence and convince others (see Chapters 2 and 3). It was not just a matter of *what* was known, but also *who* knew it. Knowledge is valueless if it remains isolated among a few experts, buried in journal articles, or fails to affect public affairs. The case raises an important question in doing biology: Who is responsible for ensuring that biological knowledge reaches the relevant audience? Also, if the information is technical, how do you make it understandable while also meaningful? Ultimately, for the vast majority of people who are not biologists, what *counts* as biological knowledge?

Carson's solution on this occasion was to write a full-length book for a public audience. Its title, *Silent Spring*, drew on an image from Mrs. Huckins: a community suddenly made silent from loss of life due to pesticides. *Silent Spring* was enormously influential. It not only raised public concerns about pesticides but also helped launch an environmental movement that still continues decades later. How did Carson solve the problem of communicating biological knowledge?

WHO NEEDED TO KNOW WHAT ABOUT DDT?

The history of one pesticide, **DDT** (dichloro-diphenyl-trichloro-ethane), illustrates how knowledge can change—and with it, social values and practices. DDT was first synthesized in 1874 and for the next several decades was just another complex molecule, of limited interest to a handful of specialists. In 1939, however, Paul Müller of Switzerland discovered that DDT could kill insects. With this new knowledge, DDT became relevant to many people in agriculture and public health.

Insects can destroy crops. They can also spread disease organisms. Mosquitoes, for instance, can carry malaria from one human or animal to another. The problem in both cases is similar, as Carson noted in her opening chapter: managing a dense population composed entirely of individuals of one species—a **monoculture**. When insect pests enter a monoculture, they can travel easily from plant to plant. As they reproduce, the rate of spreading is compounded. The pattern is similar to a disease epidemic in a human population. When a disease agent infects one human, the close proximity of humans allows it to infect many others. In both cases, halting the spread of organisms is difficult. Another problem is weeds—"pests" of another sort. The problems of weeds and pests are also familiar to homeowners trying to tend a grass lawn—another monoculture. For humans to manage crops, lawns, and the spread of disease, they must minimize the numbers of pests as much as possible; hence, pesticides.

The knowledge that DDT could kill insects changed its social value and use. During World War II, the Allied forces started using DDT to control the spread of diseases. Their success dramatically demonstrated DDT's future potential. Carson observed that "almost immediately DDT was hailed as a means of stamping out insect-borne disease and winning the farmers' war against crop destroyers overnight." The discoverer of DDT, she noted, won a Nobel Prize in 1948. Farmers, officials at the U.S. Department of Agriculture, and public health officials all saw DDT as a valuable solution to their problems.

DDT, biochemists later learned, interferes with the cell's energy-processing system. (It enters the mitochondrial membrane and destroys the energized gradient that fuels the making of ATP—see Chapter 8.) DDT and many other pesticides are thus poisons. Carson portrayed them in emotional language as "elixirs of death." Pesticides are not always lethal specifically to insects, however. The birds outside Mrs. Huckins's door had suffocated internally from DDT (at the cellular level). Such chemicals should be called "biocides," not "insecticides," Carson argued. Humans, too, could be affected in large enough doses. No wonder, Carson noted: many pesticides were derived from chemical weapons developed during World War II.

Because DDT is a complex synthetic compound, many organisms have no enzyme for breaking it down. Organisms can neither digest it as food, nor excrete it, nor destroy it—as they often do with other poisons. Rather, DDT collects in fat and liver tissues. If enough accumulates, the organism can die. If, instead, the organism is eaten, the DDT is transferred to the predator. Animals later in the food chain, such as carnivorous fish and birds of prey, tend to accumulate large amounts of DDT. The DDT concentrates in successive stages in a food chain: the effects of small amounts of DDT in the environment become amplified.

As World War II ended, Carson and others became aware of the previously unknown aspects of DDT. For them, the potential benefits of DDT were coupled with dangers to both wildlife and humans. In 1944, the American Association of Economic Entomologists thus issued a statement in an effort to amend "over-optimism and distorted impressions" about DDT. The following year, scientists wrote about the dangers of DDT in *Harper's* and *Atlantic Monthly*. In 1950, the Food and Drug

Administration announced further that it was "extremely likely the potential hazard of DDT has been underestimated." But the use of DDT and similar pesticides continued. The image of DDT as a triumph of science persisted and eclipsed other concerns.

DDT is, in some ways, unrepresentative as a pesticide, but it was central for Carson. The case of DDT also highlights the challenges facing someone who wants to inform the public about the biological implications of using any pesticide. How does this information get conveyed? Who funds publications or media presentations? Who needs to know what?

> **PROBLEM**
> Consider the task of communicating knowledge about DDT in the late 1950s to the public. Sketch the challenges involved. What information is important to convey? To whom? By what media? How might you deal with preexisting images of DDT?

A VOICE FOR SCIENCE?

Who communicates biological knowledge to the public? In this case, Rachel Carson was well positioned for the task. First, Carson had been trained in biology. She first became interested in nature during her youth, when she had wandered in the orchards and woods surrounding her home. Carson might not have become a biologist, though, had it not been for a required science course at college. Carson was inspired by her teacher, who encouraged her to pursue a career that she had not imagined (in 1926) as open to women. Carson changed majors. She went on to earn a masters degree in marine biology at Johns Hopkins University. In her thesis work, she demonstrated her meticulous skills by dissecting embryos, preparing slides, and describing in detail how a kidney first developed and then disappeared in a maturing catfish. Given her background, Carson was able to understand pesticides from a biological perspective.

Carson was also a talented writer. She could therefore interpret technical information for a general audience. Carson's writing skills were also rooted in her childhood. At age 11, she won her first writing prize—one of many over her lifetime. Rachel's mother and her teachers nurtured what they perceived as a special talent. After graduate school, Carson's opportunities took her into a career of writing about biology. At first she wrote documents for the public at the U.S. Fish and Wildlife Service (she was only the second woman hired there in a nonsecretarial position). Later, she wrote popular books about the sea, *The Sea Around Us* and *The Edge of the Sea*, which received many awards. Carson's fluid writing style earned her many readers.

Writing skills contributed to Carson's work in another significant, yet less obvious way. Carson needed to support herself while researching and writing about pesticides. Fortunately, due to the success of her books and her royalties from them, she had been able to retire from her government work. Her modest wealth allowed her to pursue and complete a project that, even with some secretarial assistance, eventually spanned four years. Later, Carson's relative financial independence would become even more important. Claims about biology can often be interpreted in the context of the organization that funds the research or publication. Because no one had sponsored Carson's work, others could not easily discount it as reflecting the bias of some specific interest group.

As reflected in her popular writing, Carson appreciated deeply the whole of nature. When she thought about pesticides, therefore, she did not view them exclusively from the perspective of agricultural yields or the eradication of disease, as farmers or public health officials did. The scope of her biological view was wider. She also considered animals in the habitat where pesticides were sprayed, the food chains of which the insects were a part, and the effect of pesticides seeping into the soil and groundwater. Carson's concerns allowed her to highlight aspects of DDT and similar pesticides that many other persons, with their own immediate concerns, did not.

Carson had not pursued a career as an ecologist or chemist. According to her credentials, then, she was something of an outsider to pesticides research. This might have affected her capacity to be an informed voice for science. Yet Carson's broader focus also allowed her to synthesize the work of many researchers. To convey her views about the causes of cancer in humans, for example, Carson drew on data from many different sources. She could thereby recognize recurring patterns that suggested a link between pesticides and cancer. While her conclusions relied on others'—some of which have since been challenged—today her broad conclusions about toxic chemicals in the environment remain confirmed. Carson was thus qualified, while differing from scientists who worked in the lab or field. Indeed, by the end of her work, Carson probably knew more about the ecological, physiological, and social aspects of pesticides than any single researcher.

Carson was guided by her expertise, but what motivates someone to distribute biological knowledge? Here, Carson's feelings about the importance of that knowledge were probably significant. In her letters discussing the proposed book, she commented how pesticides represented an "alarming threat to human welfare, and also the basic balance of nature on which human survival ultimately depends." She reported to a close friend several months later that "there would be no peace for me if I kept silent." Carson adopted the project with a sense of mission.

The effort to inform the public in this case also had political implications. Challenging the pristine cultural image of pesticides would mean taking on the entire chemicals industry. Carson would need a good dose of confidence, even if she had strong evidence to support her claims. Moreover, she was certainly not blind to assuming this role in an era when women were not yet widely respected as leaders. Carson had certainly distinguished herself during her tenure in the government, lastly as director of the Public Information Division at the U.S. Fish and Wildlife Service. Yet, as would become evident later, assuming the voice of scientific authority had significant consequences. Carson's courage and conviction were integral to her project.

THE CHALLENGE OF WRITING ABOUT PESTICIDES

The task of writing effectively about pesticides in 1958 involved many challenges. For Carson, the prospect was very different from writing about tide pools, sponges, or sea salts. Information about the dangers of pesticides would not lend itself easily to her award-winning lyrical and poetic style. In addition, the audience would be diverse, from the general public to farmers and government administrators. Finally, the topic was highly political. This would be no mere "popularization" like her earlier books.

FIGURE 17.2 An illustration opening a chapter of *Silent Spring*. Was it appropriate to couple such images with the scientific information Carson presented to the public?
Source: Illustration from *Silent Spring* by Rachel Carson.

Carson might easily have provoked public opinion by writing an inflammatory book criticizing pesticides without reference to scientific studies. Alternatively, she might have written dispassionately about the scientific details of pesticides. But Carson did neither. First, she wanted her readers to view pesticides as many biologists did. Given cultural prejudices, this would not be easy. It would not suffice merely to present information that was accurate and well documented. Carson would also have to *convince* others that the information was sound.

Carson was thus aware that her research had to be meticulous. In 1958, for example, she wrote to a former colleague asking about the decline of bird populations. But, she added, information would have to "hold up under fire." She used only claims supported by independent studies. Eventually, Carson devoted a full one-seventh of her book to documenting her sources.

At the same time, Carson chose not to write a strictly technical document. She interpreted the information in vivid images. She used such words and phrases as "sinister," "evil," "grim specter," and "ruthless power" alongside statistics or details about the ecological effects of pesticides. In addition, each chapter began with a hand-drawn and often romanticized illustration (see Figure 17.2). The technical information appeared in a primarily nontechnical context.

The sense of reverence for life, which permeated Carson's earlier writing, also guided her writing in *Silent Spring*. Rather than revel in the wonders of life, though, she portrayed the tragedy of loss of life. She cast human intervention in nature and threats to life as dangers to prevent. She selected information about the health effects of pesticide residues on food, for example, or the deaths of fish or small mammals to illustrate how respect for life had been violated. "By acquiescing in an act that can cause...suffering to a living creature," she asked her readers, "who among us is not diminished as a human being?"

The project of writing about pesticides also had a personal dimension. Ironically, Carson was diagnosed with breast cancer in 1957, just before starting work on a book that argued that the uncontrolled release of chemicals in the environment caused cancer. Carson continued treatment for her cancer through 1958. At the end of the year, her mother died. While such a loss would be difficult for almost anyone, Carson, who had been extremely close to her mother all her life, was deeply affected. Around the same time, Carson's sister died, and she assumed the additional responsibility of raising her nephew. Carson's cancer did not abate, and in 1960 she had a radical mastectomy. Carson had settled into a small community on the coast of Maine, where she became very close friends with one neighbor. Given the several tragedies that Carson experienced, we can only imagine the importance of the personal support that Carson received while writing *Silent Spring*.

FUELING AN ENVIRONMENTAL ETHOS

Carson's argument about pesticides was both specific and general. In her specific argument, she focused on how individuals use or apply chemical pesticides. Carson did not argue against all pesticides. She suggested that selective use of chemicals might be appropriate, when and where the negative effects could be controlled. Only the indiscriminate use of especially powerful pesticides posed problems.

Carson did not make her specific criticism without offering an alternative. She was well aware, for example, that arsenic compounds and other poisons used earlier had been far worse than DDT. Carson advocated instead natural chemicals and **biological control**. She urged farmers to use an understanding of nature to control nature. They should find plants that already produced chemicals that deterred insects from eating them. They should introduce insect predators, diseases, or parasites, or promote the conditions under which they would thrive. In Carson's view, the pest problem was solvable. But farmers must first learn to respect the many interactions in nature. The solution for Carson was primarily biological, not chemical.

Carson's message about biological control was lost on many readers. They may well have been responding to her more general argument and the imagery associated with it. While talking ostensibly about pesticides, Carson drew on metaphors about the balance of nature, the integrity of its interactions, and control of nature. She highlighted very specific connections. She linked the dwindling number of songbirds, for example, to human decision making, not just to pesticides. Likewise, from the failures of pesticides to control Japanese beetles and fire ants, she drew an explicit moral about human attitudes towards controlling nature. Ultimately, *Silent Spring* was not just about pesticides.

For Carson, on a more general level, humans had not considered the complex interactions of living systems. Nature had a certain integrity or balance that was disrupted when humans introduced their synthetic chemicals. Carson viewed nature holistically, as a system. She saw that an influence in one part thus had the potential to upset the system disproportionately throughout (even if at first the system could manage the slight perturbations). The fundamental problem with pesticides, Carson argued, was not simply their undesirable effects. Rather, it was human arrogance in striving to control nature. Here, Carson stepped well beyond the biology of pesticides. She critiqued public attitudes about the environment revealed by the social use of pesticides.

This larger theme formed a framework for the information about pesticides and dominated the structure, language, and examples in *Silent Spring*. The opening epigram (quoting another author) epitomized the message for many readers:

> I am pessimistic about the human race because it is too ingenious for its own good. Our approach to nature is to beat it into submission. We would stand a better chance of survival if we accommodated ourselves to this planet and viewed it appreciatively instead of skeptically and dictatorially.

Carson's book was a plea for recognizing the complexity of nature and its apparent fragility—vividly illustrated in the history of (mis)using chemical pesticides. She wanted humans to reassess their technologies and their relationship with nature.

Given the recurrent images about the control of nature, it may be somewhat ironic that Carson's alternative to chemical pesticides was itself another form of control: biological control. Yet readers perceived the book as expressing an emerging environmental way of thinking. One book reviewer may have captured the view of many readers when he wrote: "It is a devastating, heavily documented, relentless attack upon human carelessness, greed, and responsibility." Many readers did not separate Carson's specific message about pesticides from her general views about the environment, the economy, and ethics.

PROBLEM

Consider the relationship between the specific problem of pesticide use and the general problem of control of nature. Does either one necessarily imply the other? How would different readers interpret Carson's claims, based on how they perceived the problem on each of these two levels?

THE STORM FROM *SILENT SPRING*

The publication of *Silent Spring* in 1961 generated intense controversy. Chemical manufacturers responded to Carson's book in a way that reflected their interests. They said its conclusions were flawed, though they rarely addressed the results of specific studies that Carson cited. They suggested that the agricultural system and national economy would be crippled without pesticides of any kind. At the time, the United States was in a "Cold War" with the Soviet Union, and the industry cast the book in a way that drew on public fears of communism.

You might expect that a news magazine would be more likely to be objective. *Time*, however, informed readers of Carson's "oversimplifications and downright errors." Scientists and other technically informed persons, it reported, considered the book "unfair, one-sided, and hysterically overemphatic. Many of the scary generalizations—and there are many—are patently unsound." *Time*'s review praised

Carson's motives while trivializing her work: "Many scientists sympathize with Miss Carson's love of wildlife, and even with her mystical attachment to the balance of nature. But they fear that her emotional and inaccurate outburst in *Silent Spring* may do harm by alarming the nontechnical public, while doing no good for the things that she loves." Here, *Time* made a judgment on behalf of the reader about the relative authority of Carson versus others with technical credentials. What would the readers of *Time* "know"—or think they knew? How many of them do you think would have read Carson's work itself?

Most importantly, perhaps, critics sought to discredit Carson herself. They portrayed her—significantly—as a woman, "Miss" Carson, swayed by emotion: she was a "bird lover," a "cat lover," a "fish lover," a "priestess of nature"; she was a "hysterical" woman. Such epithets were misleading caricatures, of course. Yet given the cultural images of women at the time, they could raise substantial doubts about Carson's credibility. The implication was that someone who might see the emotional dimension of information could only be wrong scientifically. They also implied that a woman could be neither a spokesperson for nor authority about science. These critics sought to remove Carson's voice from the debate. Imagine how someone whose ability to present evidence objectively is itself at issue could deal with such criticism.

The book also elicited extensive public reaction. Readers expressed themselves in a "tidal wave" of letters to Carson and to newspapers, Congressmen, and government agencies. Their outrage about health and nature largely reflected the emotional dimension of Carson's writing. The book became the subject of many newspaper

J. W. Taylor in *Punch*. Copyright © Punch Publications.

"This is the dog that bit the cat that killed the rat that ate the malt that came from the grain that Jack sprayed"

FIGURE 17.3 Cartoon inspired by Carson's work. *Source:* J. W. Taylor in *Punch*. Copyright © Punch Publications.

editorials, columns, and political cartoons (Figure 17.3). How much of the response, do you suppose, was based on careful evaluation of the scientific evidence?

After *Silent Spring*, President John F. Kennedy asked his scientific advisors to examine Carson's claims. They largely confirmed her conclusions, giving them wider public authority. In the decade that followed, the U.S. Congress passed landmark legislation to regulate the use of pesticides (and to ban some, like DDT). New laws also set standards for clean air and clean water, protected endangered species, and established the Environmental Protection Agency. A mere three decades later, *Silent Spring* would be ranked as one of the 25 most influential books in human history, along with the *Bible*, Shakespeare's works, and Darwin's *On the Origin of Species*. Carson's effort to inform the public about the biology of pesticides had transformed society.

Carson would reflect much later that "some awareness of this problem has been in the air, but the ideas had to be crystallized, the facts had to be brought together in one place." Carson clearly tapped into environmental sentiments that were already emerging or latent in the populace. Still, many biologists had tried during the 1950s to alert the public—and had failed. The effects of DDT had not changed since the mid-1940s, when the earliest articles by entomologists and government officials appeared. Carson's influence was unique.

An ecologist, Murray Bookchin, published a book of similar if not wider scope a mere six months prior to *Silent Spring*. But Bookchin admitted a decade later that Carson's "superb prose" had been able to captivate a large audience in a way that he had not. For this and other reasons, Carson's book was a seed crystal and a catalyst. The image of a "silent spring" served, like a flag on a battlefield, as a rallying point for what soon became a popular environmental movement. It was a new emblem, as powerful as the images of aldrin and DDT had once been for the promise of chemical technology.

☐ *EPILOGUE*

"The" problem of pesticides may be viewed in many ways. Carson chose to focus specifically on how certain chemical pesticides were used and generally on how humans viewed nature and tended to control it. Carson's alternative was biological control.

Other views are possible. The problems posed by DDT, for example, are quite specific. Not all pesticides persist in the environment for long periods after spraying, thereby endangering other local wildlife. Nor do they all concentrate in the food chain. This occurs only when the compound is stored in some tissue and the tissue is eaten by the predator. In addition, DDT affects many organisms, not just one particular group of insects. Accordingly, many chemists since *Silent Spring* have searched for alternative synthetic chemicals that are biodegradable and species–specific.

Alternatively, you might note that we use pesticides only when plants cannot resist insect attacks themselves. Some individual plants are more susceptible than others in the same species. The prospective solution given this view is more long term: breed more resistant strains of crops. Some agronomists have considered this as another partial solution. They also recognize, however, that pests can evolve as well. Any new capacity to resist insects may be only temporary (see Chapter 13).

Finally, you could see the need for pesticides in terms of crop husbandry. That is, agriculture is often based on growing a uniform stand of one crop. This monoculture style of farming makes sowing and harvesting easier, but it also allows pests to spread rapidly. The vulnerability of monocultures to pests, recall, is one reason pest control is necessary. In many nonindustrialized nations, though, crops are often mixed on the same land and harvested separately. Where there is no monoculture, pests pose fewer problems. Thus, to solve the problem of pesticides, you would *dis*-solve the need for them.

> **PROBLEM**
> **Identify whether each of these four alternatives solves both the specific problem and the general problem that Carson introduced in *Silent Spring*.**

On the twenty-fifth anniversary of the publication of *Silent Spring*, in 1987, two friends and colleagues of Carson assessed the status of pesticides in the United States. Pesticide use in agriculture had been strongly curtailed. But groundwater was still being contaminated by chemical runoff. One study sponsored by the National Academy of Sciences compared sources of pesticides and herbicides. Average use on farmland was less than 2 pounds per acre. Use on residential lawns, however, where homeowners applied them to control weeds, averaged 10 pounds per acre. While *Silent Spring* had ushered in major changes, the question of who knew about the biology of pesticides and who did not was still relevant. Vestiges of the challenge that Carson faced in informing the public still seem to linger.

QUESTIONS AND ACTIVITIES

1. What does this case show about the following aspects of doing biology?
   — experts versus nonexperts
   — the individual versus collective nature of science
   — the role of individual motivation
   — the interaction between emotion, values, and research
   — communication and writing skills
   — the role of personal background in interpreting a scientific problem

2. Reconsider the problems about public communication of scientific information that you identified in the problem on page 188, now knowing the history of *Silent Spring*. How would you address the remaining problem of informing residential users about pesticides?

3. How was Carson's use of emotive images and language appropriate or inappropriate in informing the public?

4. Discuss whether Rachel Carson, given her unique position and abilities, had a duty or moral responsibility to write about pesticides. More generally, to what degree can society justifiably expect an individual biologist to communicate the results of his or her research to the public when the findings have broad social significance?

5. Devise a strategy of informing the public that might have won support from the chemical industry by drawing on their interests.

6. In 1962, a television news documentary presented interviews with Carson along with industry and government officials who disagreed with her. What criteria would the typical viewer have used to assess the relative credibility of each speaker? How would a news reporter have been able to assess their credibility? More broadly, how does someone who is not an expert know who is (see also Chapters 12 and 16)? Discuss how you might organize public decision making on scientific issues to take advantage of expertise while controlling bias.

7. Comment on the claim in Carson's epigram that we should view nature "appreciatively" rather than "skeptically and dictatorially"? In what ways, if any, is ecological knowledge needed to make this value judgment? What is the relationship between science and values in this case?

SUGGESTED READING AND VIEWING

Brooks, P. 1972. *The House of Life: Rachel Carson at Work*. Boston: Houghton Mifflin.

Carson, R. 1961. *Silent Spring*. Boston: Houghton Mifflin. Reprinted 1991.

McCay, M. A. 1993. *Rachel Carson*. New York: Twayne Publishers.

McKibben, B. 1989. *The End of Nature*. New York: Anchor Books.

McPhee, J. 1989. *The Control of Nature*. New York: Noonday Press.

"Rachel Carson's *Silent Spring*." 1994. From *The American Experience* series. WGBH (Boston).

Sources of Quotations

Preface
H. A. Krebs, "The History of the Tricarboxylic Acid Cycle," *Perspectives in Biology and Medicine* (1970) 14: 154–170.

Chapter 1
J. R. G. Turner, "Henry Bernard Davis Kettlewell." In C. G. Gillespie, ed., *The Dictionary of Scientific Biography*, vol. 17, supplement 2, pp. 469–471.

Chapter 3
E. B. Wilson, *The Cell in Development and Heredity*, 3rd ed. New York: Macmillan, 1925, p. 739.

Chapter 6
R. Dubos, *The Professor, The Institute, and DNA*. New York: Rockefeller University Press, 1976, pp. 91, 140.
E. P. Fischer and C. Lipson, *Thinking About Science: Max Delbruck and the Origins of Molecular Biology*. New York: Norton, 1988, pp. 151, 152.

Chapter 8
Efraim Racker and Thomas E. Conover, "Multiple Coupling Factors in Oxidative Phosphorylation," *Federation Proceedings* (1963) 22: 1088.
Alexander Tzagoloff, *The Mitochondrion*. New York: Plenum Press, 1982, p. 131.
E. C. Slater, "Mechanism of Energy Conservation in Mitochondrial Oxido-reductions." In J. M. Tager, S. Papa, E. Quagliarello, and E. C. Slater, eds., *Regulation of Metabolic Processes in Mitochondria*. Amsterdam: Elsevier, p. 174.
Harold Franklin in a memorial note.

Chapter 9
W. B. Cannon, *The Way of an Investigator: A Scientist's Experiences in Medical Research*. New York: W. W. Norton, 1945, pp. 27, 94.

Chapter 10
H. Selye, *The Stress of Life*. New York: McGraw-Hill, 1956, pp. 16, 28.

Chapter 14
N. Tinbergen, "The Curious Behavior of Sticklebacks," *Scientific American* (1952) 187(6): 22.

Chapter 15
R. W. Clark, *JBS: The Life and Work of J. B. S. Haldane*. New York: Coward-McCann, 1969, p. 68.

Chapter 17
Paul Brooks, *Rachel Carson: The House of Life*. Boston: Houghton-Mifflin, 1972, pp. 228, 232.
Loren Eiseley, review of *Silent Spring*, *Saturday Review* 45 (September 29, 1962), p. 18.

Chapter Index of Themes
in *Doing Biology*

Evidence and Scientific Thinking

analogies 3, 13
assumptions 3, 15
controlled experiment 1, 6, 7, 9, 10, 11, 12
    causation versus correlation 10, 11
    double-blind studies 12
    field experiments 1, 11
models 8, 10, 15
multiple types of evidence 1, 3, 6
nonexperimental methods 4, 14
posing problems 3, 5, 10, 13, 14, 17
relationship between theory and experiment 1, 9, 16
replication of experiments 1, 3, 6, 10, 12
role of false hypotheses 3, 7, 8
role of theoretical perspective in interpreting observations 1, 3, 5, 10, 11, 12, 13, 14, 17
skepticism 5, 6, 10, 12, 16

The Process and Nature of Science

chance or accident 5, 6, 11
creativity 7
incremental nature of discovery and acceptance of theories 3, 7, 9, 11
personal motivation 7, 10, 17
personality 2, 6, 10
reconceptualizations 2, 8, 11
relationship between basic and applied research 6
revision of theories and problems 2, 3, 4, 8, 9, 13, 14, 15, 16
scientific instruments and methods 7, 14

The Scientific Profession and Community

collaboration and collective work 6, 8, 11, 17
communication 2, 6, 11, 13
disciplinary boundaries and interrelationship of disciplines 2, 3, 5, 8, 9, 13, 14, 16
persuasion of colleagues 2, 3, 6, 10
    areas of expertise 6, 12, 16, 17
    burden of proof 6, 12, 17
professional training 4, 7, 17
research ethics 9, 11, 12
resolution of disagreement 3, 4, 8, 16

The Social and Cultural Contexts of Science

costs and funding research 5, 11, 12
cultural context of scientific norms 12
public understanding of science 9, 12, 17
social responsibility of scientists 9, 17

Index